Environmental Innovation and Firm Performance

Environmental Innovation and Firm Performance

A Natural Resource-Based View

Javier Amores Salvadó, Gregorio Martín de Castro,
José Emilio Navas López and Miriam Delgado Verde
Universidad Complutense de Madrid, Spain

© Javier Amores Salvadó, Gregorio Martín de Castro, José Emilio Navas López and Miriam Delgado Verde 2013
Foreword © Stuart Hart 2013

All rights reserved. No reproduction, copy or transmission of this publication may be made without written permission.

No portion of this publication may be reproduced, copied or transmitted save with written permission or in accordance with the provisions of the Copyright, Designs and Patents Act 1988, or under the terms of any licence permitting limited copying issued by the Copyright Licensing Agency, Saffron House, 6–10 Kirby Street, London EC1N 8TS.

Any person who does any unauthorized act in relation to this publication may be liable to criminal prosecution and civil claims for damages.

The authors have asserted their rights to be identified as the authors of this work in accordance with the Copyright, Designs and Patents Act 1988.

First published 2013 by
PALGRAVE MACMILLAN

Palgrave Macmillan in the UK is an imprint of Macmillan Publishers Limited, registered in England, company number 785998, of Houndmills, Basingstoke, Hampshire RG21 6XS.

Palgrave Macmillan in the US is a division of St Martin's Press LLC, 175 Fifth Avenue, New York, NY 10010.

Palgrave Macmillan is the global academic imprint of the above companies and has companies and representatives throughout the world.

Palgrave® and Macmillan® are registered trademarks in the United States, the United Kingdom, Europe and other countries.

ISBN 978–0–230–36347–2

This book is printed on paper suitable for recycling and made from fully managed and sustained forest sources. Logging, pulping and manufacturing processes are expected to conform to the environmental regulations of the country of origin.

A catalogue record for this book is available from the British Library.

A catalog record for this book is available from the Library of Congress.

10 9 8 7 6 5 4 3 2 1
22 21 20 19 18 17 16 15 14 13

Contents

List of Figures and Tables viii

Foreword x
Stuart L. Hart

Acknowledgements xii

1 Theoretical Framework 1
  1.1 Introduction 1
  1.2 Environmental concern and economic activity 2
      1.2.1 From ignorance to awareness 3
      1.2.2 From awareness to conflict 4
      1.2.3 From conflict to solutions 6
      1.2.4 One step beyond 8
      1.2.5 The role of the Natural Resource Based View in the sustainability debate 9
  1.3 Natural Resource Based View 10
      1.3.1 Back to the roots: The resource based view 10
      1.3.2 Steps to corporate sustainability 18
      1.3.3 Defining the scope 22

2 The Nature of Environmental Product Innovation and Green Image 24
  2.1 Introduction 24
  2.2 Innovation: Approaches to the term 25
  2.3 Socio-economic innovations 29
      2.3.1 Environmental innovation. Definition, determinants and types 32
      2.3.2 Environmental process innovations 40
      2.3.3 Environmental organizational innovations 45
      2.3.4 Environmental product innovations 48
      2.3.5 Environmental tools as supporting factors 51
      2.3.6 Other approaches to environmental innovation 58

2.4 Green corporate reputation and image ... 64
  2.4.1 Corporate reputation. Review, proposed definition and characterization from a management perspective ... 65
  2.4.2 Corporate reputation, corporate image and corporate identity: Exploring the 'terminological jungle' ... 72
  2.4.3 Corporate reputation: Main features ... 74
  2.4.4 Social and green corporate reputation ... 75

3 **Research Model** ... 79
 3.1 Introduction ... 79
 3.2 Individual effects ... 80
  3.2.1 Environmental product innovation and firms' performance ... 80
  3.2.2 Green corporate image and firms' performance ... 82
 3.3 Environmental product innovation and green corporate image: Their joint effect on firms' performance ... 85

4 **Methodology** ... 88
 4.1 Introduction ... 88
 4.2 Environmental innovation: Previous empirical studies ... 88
 4.3 Measurement of variables ... 104
  4.3.1 Variables development: Measurement scales ... 104
  4.3.2 Environmental product innovation ... 104
  4.3.3 Green image ... 105
  4.3.4 Firm performance ... 106
  4.3.5 Control variable ... 110
 4.4 Information sources and data gathering ... 110
 4.5 Sample characteristics and statistical representativeness ... 112

5 **Research Results** ... 120
 5.1 Introduction ... 120
 5.2 The Spanish metal sector: Overview of environmental product innovation and green image initiatives ... 120

	5.3	Exploratory analysis	124
		5.3.1 Introduction	124
		5.3.2 Environmental product innovation	125
		5.3.3 Green image	127
		5.3.4 Firm performance	128
	5.4	Exploring the relationship between environmental product innovation, green image and firm performance	128
6	**Conclusions**	132
	6.1	Introduction	132
	6.2	The evolving nature of the Natural Resource Based View	133
	6.3	Implications	135
	6.4	Limitations and future research lines	136

Annexes   139

Notes   147

References 148

Index   161

Figures and Tables

Figures

1.1	RBV's 'spin-offs' and theoretical developments	13
1.2	NRBV: Conceptual framework	18
1.3	Environmental innovation and sustainable strategies	20
2.1	Environmental innovation types	39
2.2	Environmental product innovation dimensions	51
2.3	Corporate reputation, corporate image, corporate identity	73
2.4	Corporate reputation: Main strategic characteristics	74
3.1	Research model: Individual and mediated effects	85
5.1	Environmental product innovation measures. Distribution in use	121
5.2	Green image measures. Distribution in use	122
5.3	Firm profiles for environmental product innovation	123
5.4	Firm profiles for green image	123

Tables

2.1	Innovation definitions	25
2.2	Innovation orientations	28
2.3	Socio-economic innovations	30
2.4	Environmental innovation definitions	34
2.5	Environmental process innovation definitions	41
2.6	Environmental product innovation definitions	49
2.7	The treatment of corporate image and reputation from different disciplines	67
2.8	Reputation as 'collective awareness'	68
2.9	Corporate reputation as assessment	69
2.10	Reputation as an asset/resource	71
2.11	Dimensions of corporate reputation	76
4.1	Previous empirical studies on environmental innovation and environmental management	90
4.2	Environmental product innovation indicators	105

4.3	Green image indicators	105
4.4	Studies that measure firm performance from environmental approach	106
4.5	Firm performance indicators	110
4.6	Control variable	110
4.7	Questionnaire structure – questions directed to innovation/environmental managers	112
4.8	Questionnaire structure – questions directed to general or financial managers	112
4.9	Industrial activities included in the research	113
4.10	Industries' turnover	115
4.11	Number of firms	116
4.12	Consumptions (year 2009)	117
4.13	Research facts and figures	119
4.14	Sample statistical representativity	119
5.1	Correlations matrix	125
5.2	Main components analysis of environmental product innovation	126
5.3	Main component analysis of green image	127
5.4	Main component analysis of firm performance	129
5.5	Regression analysis results	130

Foreword

Stuart L. Hart
Cornell University

I first wrote the draft of what would ultimately be published as "A Natural-Resource-Based View of the Firm" (Hart, 1995) nearly 20 years ago. Even then, the pollution prevention agenda was well established and 'eco-efficiency' was *de rigeur* among progressive corporations. My intent in crafting the piece was to move the emerging field of sustainable business beyond the simple logic of waste, cost and risk reduction. I sought to embed the environmental challenge more centrally in the conversation about strategy and strategic management. Accordingly, I spent considerable time conceptualizing the potential sources of sustained competitive advantage to be gained through broader product stewardship strategies, and the emerging domain of 'sustainable development'. I was especially interested in how the challenge of global sustainability might become a key driver of innovation, both in the established markets of the developed world and in the emerging markets of the developing world. Indeed, most of my subsequent work has been devoted to articulating the strategic logic for moving 'beyond greening' through leapfrog technological innovation and including the underserved at the base of the world income pyramid (e.g. Hart, 1997; Prahalad and Hart, 2002; Hart, 2005).

Now, nearly 20 years later, extensive empirical research has tested a number of the propositions associated with the Natural Resource Based View (NRBV). As my colleague Glen Dowell and I have pointed out, however, this work has focused primarily on uncovering the links between pollution prevention and firm profitability, while the other areas of the NRBV have been explored to a much less thorough degree (Hart and Dowell, 2010). Indeed, while significant conceptual and practitioner literatures have emerged on innovations such as green supply chains, life cycle design, clean technology, social entrepreneurship and base of the pyramid strategies, there is still little in the way of empirical work focused on these topics.

It is for this reason that the publication of *Environmental Innovation and Firm Performance* by Javier Amores Salvadó and his colleagues is so significant. The book purposefully moves beyond the eco-efficiency agenda to focus on the impacts of environmental product innovation and green corporate image. And by examining empirically the connection between these two strategies on firm performance, they provide important insights into how internal (life cycle design) and external (image and reputation) innovation strategies serve to mediate and (potentially) reinforce one another.

The authors not only effectively apply the NRBV but also (and perhaps more importantly) significantly advance the theory. As a consequence, this book marks an important advance in the field of sustainable enterprise.

Acknowledgements

We wish to acknowledge all the people who have believed in our work and have made it possible. In particular, we want to thank professors Aragón-Correa, González-Benito and Molina-Azorin for sharing their knowledge with us and giving support and valuable advice when we needed it.

We would also like to thank all the institutions whose support have made this investigation possible, particularly the Ikujiro Nonaka Research Centre for Knowledge and Innovation and the Spanish Ministry of Science and Innovation (project ECO2009-12405).

Finally, we would like to express our gratitude to all the Spanish managers and organizations that have devoted their time and effort to this research. To all of them, thank you very much.

1
Theoretical Framework

1.1 Introduction

Although there are many voices that highlight the importance of taking into consideration the restrictions imposed by the environment on the development of economic activity, at this moment, progress made in this regard is still scant. Nevertheless, sooner rather than later, companies will have to deal with a scenario of scarce natural resources where environmental orientation will be a first-order competitive element.

This book is the result of this concern. Its main aim is to show the theoretical foundations that sustain the economic logic behind the incorporation of environmental arguments to business activities. In order to shed some light on the potential benefits of environmental practices, we analyse the effect of two specific dimensions of product stewardship strategies (Hart, 1995) on business results. Specifically, we focus on the role of environmental product innovation and green image as two of the main driving forces of the relationship between the firm and the entire value chain made up of customers, suppliers and other stakeholders.

Thus, detailed analysis of firms' innovative environmental practices and their commitment towards an image of environmental friendliness provide us with a fundamental knowledge base that, in future studies, should lead to the identification of those specific environmental capabilities that are needed to successfully tackle the sustainability social and economic challenge.

Therefore, in order to appropriately address the questions raised above, from a management approach, we focus on two main theoretical streams.

- The Natural Resource-Based View (Hart, 1995): The Natural Resource-Based View (NRBV) is the cornerstone of our argument. Taking the Resource Based View as a reference, it attempts to answer to the challenge of sustainable development. From this view, current economic patterns are not environmentally sustainable and firms, in order to be competitive in tomorrow's economy, must develop the necessary resources and capabilities that enable them to adapt to the constraints imposed by the natural environment.
- Resource Based View of the firm (Wernerfelt, 1984; Barney, 1991, 2001; Peteraf, 1993). In line with what we have stated above, on the one hand, the incorporation of environmental arguments to processes, products and organizational modes to the firm, and on the other hand, competitive advantage achievement, require the development of a number of specific resources and capabilities.

Besides its importance as the 'ecological evolution' of the Resource Based View theory of the firm, the NRBV also plays a major role in the evolution of the link between the firm and the natural environment. Through the analysis of the different stages into the environmental management literature, we will appreciate the real value of this theory as it is the one that puts together the ecological concern and the competitive advantage achievement.

After reviewing and assessing the role of the NRBV in the environmental management literature, we will develop in detail its Resource Based View origins. Taking as the key reference the NRBV, in the following lines, we will proceed to analyse its fundamentals, general scope and the specific aspects of the product stewardship strategy to which we have devoted a significant part of our work.

1.2 Environmental concern and economic activity

We have to look back to the last century to differentiate between the three different stages or steps in the environmental concern/economic activity relationship. To do this analysis we will follow most of the assumptions proposed by Colby's (Colby, 1991) and Del Brio's (Del Brio and Junquera, 2001) classifications. According to their classifications, the first stage, where the environmental issues were almost completely absent from the economic debate, ranges from the beginning of industrialization to the early 1970s. In that period, firms were only worried about their competitive markets. The environmental issues were not considered at all in their strategic agenda, and the dominant theoretical approach was the Frontier Economics paradigm. The second stage in this theoretical evolution is characterized by the emergence of conflict between Deep Ecology and Environmental Protection principles. While the former approach is claiming humans' subordination to nature, the last is trying to find the way to relieve the damage caused by firms' industrial activities. The third stage is dominated by the Resource Management paradigm. Its advocates are aware of the fact that the solution to the environmental problem is not only in the implementation of Environmental Protection measures. At this time the main concern is how to set the 'right' prices for environmental goods. In the last stage the Eco-development paradigm emerges. This approach finds that Resource Management's solutions do not face the real problem and instead more environmentally sound measures are needed. Scholars of this approach support prevention measures as well as changes in values and lifestyles.

As we will see, the Natural Resource-Based View plays a major role in this evolution. Its main assumptions serve to connect some elements of the neoclassical logic such as cost reductions and competitive advantage achievement to more ecologically and socially sound initiatives without renouncing to profit maximization. Next, we will detail the main features of every stage as well as the role of the NRBV in this evolution.

1.2.1 From ignorance to awareness

During the first decades of industrialization, in parallel with the scientific and technological knowledge advances that were emerging, a large amount of waste materials started to be thrown into the natural environment. At that time, it was generally accepted by economists that nature had the capacity to absorb and tolerate whatever harm or injury was caused by the economic activity. Furthermore, according to that conception, natural resources were unlimited and were not affected by human activity due to its regenerative power. While in the most advanced industrialized countries people's standard of living was significantly improved, the natural environment was seriously altered.

This approach is in line with what Professor Colby calls the Frontier Economics paradigm. According to that view, nature was an infinite supply of physical resources and the economic thinking does not have to be worried about the role of nature. Furthermore, the main issues to be studied and debated deal more with the dichotomy resource allocation versus distribution than with the depletion of natural resources.

Although this approach could be understandable in the early days of industrialization, this conception of the relationship between the natural environment and the economic activity was challenged by the evidence. Thus, during the 1960s environmental pollution was intensified and industrial activity started to cause damage to the health and well-being of the people, causing climate changes, atmospheric alterations, air and water pollution and also hazardous chemical waste.

As a response, environmental awareness started to grow in the industrialized societies arguing that previous conceptions based on the indefinite economic growth were unrealistic and, what is more important, unsustainable. Proof of this socio-economic change was the United Nations' General Assembly Resolution of December 3rd in 1968 convening a United Nations Conference on the Human Environment[1] noting 'the continuing and accelerating impairment of the quality of the human environment and its consequent effects on the condition of man, his physical, mental and social well being, his dignity and his enjoyment of basic human rights, in developing as well as developed countries'. This was the first international document and institutional effort recognizing that 'the relationship between man and his environment is undergoing profound changes in the wake of modern scientific and technological developments'.

1.2.2 From awareness to conflict

In 1972 some important events occurred in the environmental field.

On the one hand, the first United Nations Conference on the Human Environment was held. One of the outcomes of this conference was the establishment of the United Nations Environment Programme (UNEP), the leading programme within the United Nations working on environmental issues.

On the other hand, the book *The Limits to Growth* (Meadows et al., 1972, in Colby, 1991) was released. Recognized as one of the main landmarks of the environmental concern, it had a great influence in that time, and its main contribution was to provide an early warning of the potential problems derived from the trade-off between environmental protection and uncontrolled economic growth.

These two events are interrelated as the final statement of the United Nations Conference on the Human Environment claims that the searching of economic growth as the primary goal does not necessarily lead to greater social and living standards as natural capital is becoming more and more scarce and is subject to physical limits.

Since that moment, firms' and institutional pioneering initiatives in the field timidly started to emerge. Although the environmental issues were already in the debate, the mentioned initiatives did not go far beyond recognizing the problem and being aware of some of its consequences. At most, firms started to control (instead of prevent) their polluting emissions through the utilization of emission filters and end of pipe controls.

Behind this apparent inaction and the absence of more imaginative initiatives in regard to the environmental prevention was the conflict between two separate conceptions of the human–nature relationship: on the one hand the Deep Ecology approach, which is totally opposed to the Frontier Economics paradigm, and on the other hand the Environmental Protection view, which emphasizes the necessity to make compromises between two interconnected but sometimes opposing realities such as the economic activity and respect for the environment (Colby, 1991).

Deep Ecology advocates facing this conflict with a 'take it or leave it' approach, arguing that most technological advances usually lead to more environmental problems and cannot be categorized as progress. Consequently, it would be desirable to go back to pre-industrial and rural lifestyles in order to live in harmony with the natural environment. Conversely, the Environmental Protection response to this problem was to apply a defensive strategy focused on repairing the damage caused.

Therefore, if the economic growth and the environmental improvement were understood as two conflicting realities, should we give up one for the other?

1.2.3 From conflict to solutions

The answer to that question came with The Report of the World Commission on Environment and Development (also known as the Brundtland Commission). The report, released in 1987 and entitled Our Common Future from One Earth to One World[2], marked a point of inflexion in the way firms and institutions face the environmental degradation of the planet. By popularizing the concept of sustainable development, the commission tried to overcome the limitations of previous approaches, arguing that although there is no doubt about the successes of the industrialized societies in terms of human life expectancy, education, infant mortality or global food production, it is also true that this development have altered the planet, threatened the lives of many species and also the basis of human existence. In this sense, to mention only a few of these negative impacts, it is worth noting that the forest is still being destroyed in many countries; the burning of fossil fuels is generating carbon dioxide and causing global warming, and the desertification is, worryingly, spreading, affecting increasingly larger areas.

In other words, it must be possible to reconcile the economic growth with the preservation of the natural environment, and the natural resources must be consumed at a rate that allows the ecosystems to naturally regenerate themselves. In sum, 'meet out current needs without compromising the ability of future generations to meet their needs'.

Focusing more in the sustainable development/firms' relationship, the Brundtland Report highlighted four key points to be considered. First, although the manufactured goods are increasing, such increase is not enough to cover the necessities of the developing countries, which are constantly increasing their consumption levels. Second, new technologies offer great efficiencies and the anti-pollution measures have proved to be a good alternative to increase firms' profitability and avoid environmental damage. Third, the transnational corporations have also their responsibility in the sustainable development challenge. They (together with the governments and institutions) must collaborate and assist developing countries so they can make better use of technology and try to prevent them from the same errors that industrialized countries committed in the past. Fourth, industries have a major role in providing goods and services that sometimes are essential to meet basic human needs. Therefore, they will live up to their responsibility.

As noted by Shrivastava (1995), the idea of sustainable development has also been criticized. Some argue that: although it seeks the management of global ecological resources and systems, even the researchers do not understand completely how they function; it superficially deals with indigenous people's rights to decide about their own resources, limiting also their development options; and it is contradictory as it seeks to conserve the natural environment without seriously dealing with the current uncontrolled economic growth.

Nevertheless, although sustainable development has not solved the problems highlighted above, it has proved to be a good starting point, particularly for firms, which, drawing on sustainable development principles, have started to get more involved in the environmental field, putting into practice several initiatives like pollution prevention, waste minimization and clean technologies among others.

The Brundtland Report is very close to another environmental management paradigm, namely the Resource Management paradigm. It tries to apply the neoclassical vision but taking into consideration the fact that natural resources have also to be managed. Economic growth is still the major goal but it has to be achieved by following a 'sustainable path'. Under this paradigm the consumption levels in the developed countries as well as the uncontrolled population increases in the developing world are unaffordable in the long term, and new solutions that go beyond the environmental control are proposed. This is a case of the establishment of the 'polluter pays principle' or the tradable emissions permit, which tries to 'economize the ecology' (Colby, 1991: 204). Under this paradigm, the environmental factor has two faces: on the one hand it is the crucial factor to achieve sustainable development and on the other hand is considered as a necessary evil to live with.

We can also include in this category those authors belonging to the open-minded neoclassical environmental economics' group (Illge and Schawarze, 2009), whose members reject the introduction of fundamental changes to the economic system as well as the restrictions of material consumption. On the contrary, they also support the settlement of 'right' prices for environmental goods, which is, as we have seen, one of the main claims of the Resource Management advocates.

1.2.4 One step beyond

Although it is a step in the right direction, the above Resource Management measures are not a real challenge to firms. In fact, according to the Eco-development paradigm, what the 'polluter pays' principle is creating is a market for *bads* where the right to pollute seems to be the fundamental one. Actually, what this paradigm is claiming is the move from the polluter pays to the pollution prevention pays principle. The question is not whether the firm must repair the environmental damage, but to avoid the existence of that damage from the very beginning of the production process.

The constant trade-off between firms and the natural environment can be explained according to three basic areas (Shrivastava, 1995), namely, inputs, throughputs or production processes and outputs. Regarding the inputs, firms must pay special attention to the use and conservation of the energy. To that end, measures like product redesign, renewable materials utilization or ecologically sensitive purchasing policies are good examples of environmental commitment. Looking at the throughput system, the preventive action and the continuous improvement are the key elements to improve the efficiency, reduce the cost of production or eliminate unnecessary production steps. Finally, regarding the outputs, low consumption and long-lasting products are, together with ready-to-disassemble products, other interesting options to reduce the impact on the environment.

But firms are not operating in isolation. Sustainable development principles cover not only the environmental and economic spheres but also the social dimension of business activities. The Eco-development paradigm goes beyond the efficiency discourse and argues that it is not only efficiency but also social equity rooted in the development of new fields of knowledge like ecological engineering or industrial ecology and new social values.

The Ecological Economics scholars (Illge and Schawarze, 2009) are a good example of the defence of these statements as they also claim for the inclusion of ethical dimensions in the sustainability debate and they prefer talking about development instead of economic growth. From this approach, societal welfare is a complex reality where economic, ecological and social concerns must be satisfied. As it can be seen, this is very close to the very commonly used triple bottom line Corporate Social Responsibility approach.

1.2.5 The role of the Natural Resource Based View in the sustainability debate

The NRBV approach can be considered one of the major contributions to the field of sustainability, not only because it integrates in the same framework some of the Resource Management and Eco-development assumptions, but also because it shows how these principles can be implemented in the firms in order to achieve competitive advantages and contribute to the sustainable development goal.

As we will see in the next chapters, it incorporates some of the assumptions most commonly used by neoclassical logic when referring to competitive advantage through lower cost and staying ahead of the competition. Cost reductions can be achieved by applying pollution prevention strategies while better market positions can be achieved through product stewardship. Furthermore, NRBV pollution prevention strategies are also in line with the Eco-development paradigm and the already mentioned pollution prevention pays principle.

But the NRBV also goes beyond pollution prevention and incorporates most advanced developments from the Eco-development stream. Hart's framework claims for radical technological and social changes. On the technological side, the search for new developments aimed to mimic the processes of nature are encouraged, while on the social side, bottom of the pyramid strategies aimed to cover the needs of those collectives most disadvantaged are emphasized as well. It can be argued also that these strategies will at the end incorporate those 'at the fringe' to the mass markets through the development of disruptive technologies, which is at the same time a huge opportunity for firms and a good way of promoting social inclusion.

In sum, the NRBV fills the gap between Resource Management and Eco-development approaches and is, as we will see, the perfect framework to reconcile firms' main objectives, like competitive advantage and profit generation with environmental respect and social inclusion.

1.3 Natural Resource Based View

The NRBV (Hart, 1995) attempts to reconcile the Resource Based View (RBV) with the constraints imposed by the environment. From this approach, it is argued that the natural environment is a source of new and emerging business opportunities and firms that are able to adapt their activity to these constraints will drive the economy of the future.

A change of paradigm is necessary. The current patterns of population growth, together with the increase in industrial production and consumption levels, make the current model environmentally unsustainable. As a consequence, new environmentally friendly business strategies that improve the resources and capabilities endowments are needed and the new economic scenario is demanding other ways of competitive advantage achievement.

Nevertheless, in order to understand the key concepts of the NRBV, we will, first of all, summarize and review its theoretical basis. In other words, we will analyse briefly the RBV of the firm and its main assumptions.

1.3.1 Back to the roots: The resource based view

The RBV tries to answer the concerns of a group of scholars (Wernerfelt, 1984; Barney, 1986; Rumelt, 1991) that stressed the importance of endogenous factors in firm performance achievement.

According to this theory, competitive advantage lies within the firm and in the resources it controls or possesses. This line of reasoning was first introduced by Penrose (1959, in Mahoney and Pandian, 1992) and Learned et al. (1969, in Barney, 1991) who already spoke about the firm strengths and weaknesses highlighting the role of the resources owned by the company. However, Barney (1991) was the one that presented a robust theoretical framework on the RBV, which is nowadays considered one of the principal theories within the field of management (Acedo et al., 2006; Newbert, 2007).

The basic assumptions of this theory are:

- Firms within an industry or group may be heterogeneous with respect to the strategic resources they control.
- These resources are not perfectly mobile between firms and thus heterogeneity may persist over time.

Therefore, firms can achieve sustainable competitive advantages by implementing strategies that exploit their internal strengths, while at the same time neutralizing the potential threats emerging from the environment.

The RBV view emphasizes the role of resources and capabilities as the basis for competitive advantage. Nevertheless, in the literature, we can find slightly different approaches to the basic concepts. Thus, Wernerfelt (1984) defined resources as anything that can be considered as the strengths and weaknesses of a firm; Barney (1991) included in the concept all assets, capabilities, organizational processes, knowledge and information controlled by the firm, with the potential to increase its effectiveness and efficiency; Grant (1991) makes a difference between resources and capabilities, arguing that the resources are the inputs of the production process and basis for the capabilities; and Amit and Schoemaker (1993) stated that resources were factors endowments controlled or possessed by the firm.

According to Maijoor and Witteloostuijn (1996, in Newbert, 2008), resources are tangible or intangible assets associated to the firm in a stable way, and Barney (2001) enriched its previous definition to consider resources as tangible and intangible assets used by the firm to choose and implement its strategies.

We can also find differences among the scholars regarding the concept of capability. Itami (1987, in Amit and Shoemaker, 1993) calls them invisible assets; Amit and Schoemaker (1993) rather refer to them as firms' ability to combine and exploit resources through organizational routines in order to reach organization goals; according to Dutta et al. (2005, in Delgado, 2009) the capabilities are the efficiency with which the firm converts inputs into outputs, while Grant (1991) finds no difference between capabilities and firm abilities. In his work, Grant stressed the importance of complex capabilities, arguing that complex capabilities are a major entry barrier for the competitors, thus the firm that possesses them is in a better position to sustain competitive advantage.

In short, we can say that a resource is something that a firm possesses, it can include physical and financial assets, employees' abilities and organizational processes. However, a capability is something that the firm is able to perform or conduct and that finds its basis in the resources and routines of the organization (Hart and Dowell, 2010).

According to this theoretical stream, the firm can reach the competitive advantage if it possesses and exploits valuable and rare resources and capabilities. This way, if these resources and capabilities are also imperfect, imitable and non-substitutable the firm could sustain that competitive advantage in the time, making possible to obtain better yields in the short and long terms (Barney, 1991; Newbert, 2008).

Therefore, we can say that a resource is valuable if it reduces the production cost or increases consumer willingness to pay a higher price, and rare if it gives the firm the opportunity to sell at higher prices. Thus, the inimitability derived, for example, from the social complexity of the resource, creates the potential to sustain this advantage (Hart and Dowell, 2010).

Some scholars argue that the extent to which resources and capabilities are able to create value depends on the presence of complementary assets. In this sense, Amit and Schoemaker (1998) analysed the concept of complementarity, stressing that the strategic asset value (individually considered) rises when the value of the asset they complement also increases. In the same vein, other important contributions (Teece, 1987; Peteraf, 1993) developed the notion of co-specialized assets, which can be understood as those between whom there is a bilateral relationship of dependence in their application or deployment. This way, the combined value of firm resources and capabilities is greater than the cost of developing each one of these resources and capabilities individually.

This notion of complementarity is particularly relevant in the environmental management literature (Christmann, 2000) and though not directly addressed in this book, it is clear that the complementarities of both environmental product innovation and green image (which constitutes the key elements of out argumentation) present interesting opportunities for the firms and are a highly promising field for future research.

In spite of being considered as the dominant theoretical mainstream from the last decade of the twentieth century to the present (Acedo et al., 2006; Newbert, 2007), the RBV is a mature theory (Barney et al., 2011) that has also been criticized by some scholars due to a number of limitations such as (Priem and Butler, 2001):

- It is tautological: This means that the basic assumptions of the theory are true by definition. The value of a resource is exogenous to the theory itself and therefore this provides no argument to determine that value.
- The prescriptive character of RBV is limited because the strategic advantage-generating attributes of the resources cannot be manipulated by managers.
- Its static character, considering how organizational resources and capabilities are accumulated and used, as well as its difficulties in measurement and assessment of resources and capabilities, makes the relatively small quantity of empirical research appear in comparison with the enormous quantity of theoretical proposals that have been emerged in the last 25 years (Newbert, 2008).

All these concerns, jointly with the increasing necessity of inserting environmental and social issues in the sustainability of competitive advantages and business activities, have given way to the emergence of some other related and complementary theoretical streams or 'spin-offs' during recent years (Figure 1.1).

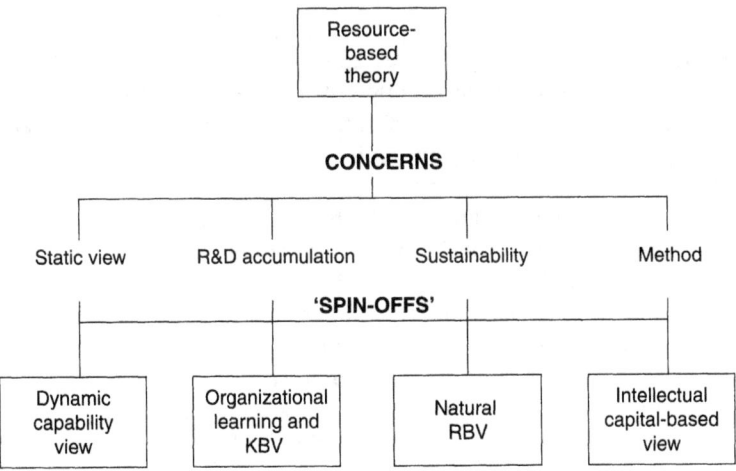

Figure 1.1 RBV's 'spin-offs' and theoretical developments
Source: Own elaboration based on Martín de Castro et al. (2011).

Thus, Dynamic Capabilities View (Teece et al., 1997; Wang and Ahmed, 2007) leads RBV from a static point of view towards a dynamic view, signalling the existence and importance of dynamic capabilities – the firm's ability to integrate, build and reconfigure internal and external competences, or the bundle of firm's resources and capabilities, to address rapidly changing environments. Dynamic capabilities like absorptive capacity or innovation capability are a major issue when dealing with turbulent and dynamic environments where firms constantly need to reconfigure and adapt their bundle of resources and capabilities to their changing environmental circumstances.

Therefore, seems clear also that the idea of change associated with the concept of sustainable development must be put into practice by companies with new capabilities that allow them to interact with interest groups so far ignored. Firms have to reconfigure and create new capabilities in order to be able to face this environmental challenge.

Nevertheless, the definition of dynamic capabilities is far from consolidated. The concept has been criticized for being vague (Kraatz and Zajac, 2001), confusing and mysterious (Winter, 2003), abstract and unworkable (Danneels, 2008) and dark and tautological (Williamson, 1999). In addition, the approach taken to address the Dynamic Capabilities has also been mixed. Thus, some authors link Dynamic Capabilities with the performance directly (Teece et al., 1997; Makadok, 2001; Zollo and Winter, 2002), for others, future economic results are dependent on the new configuration of resources and on how managers implement them (Eisenhardt and Martin, 2000; Helfat et al., 2007) and finally, there are writers who claim that between Dynamic Capabilities and performance there is an indirect relationship. This is the approach followed by Zott (2003), who considers that Dynamic Capabilities can actually change the resource base of the company, so that the new resource base will influence new market positions, which subsequently will affect performance.

Dynamic Capabilities are also very important for the NRBV. They are critical both for pollution prevention and product stewardship and also to the sustainable development capability and its relationship with the communities at the Bottom of the Pyramid (BOP). When the company creates communities in the BOP, it will not be looking for direct and immediate economic returns, but will be creating a network of relationships based on mutual trust that later and indirectly will generate yields and new market positions.

Since its origins, RBV has focused on the presence of resource capability and the achievement of sustained competitive advantage, avoiding the central issue about how and under which organizational and environmental conditions those resources and capabilities were built. Wernerfelt (2011) considers the processes of firms' resource acquisition and internal development as determinant of firm profitability. Nevertheless, in this way, since the early 1990s different proposals (Kogut and Zander, 1992; Nonaka and Takeuchi, 1995; Grant, 1996, among others) address this key issue configuring a Knowledge-based View and the parallel theoretical development of Organizational Learning Perspective. Both frameworks, jointly with the developments about Absorptive Capacity (Zhara and George, 2002) configures a current 'spin-off' and/or theoretical development of RBV that tries to explain how intangible resources, capabilities, dynamic capabilities, or knowledge is created and transferred inside and outside the firm, in order to create new knowledge and innovation, being the main vehicle in achieving competitive advantages.

These knowledge-based conceptions have also a good fit into the sustainable development literature as they are drawn on the idea that sustainable development cannot be promoted only by technological innovation (Sabau, 2010). Therefore, knowledge, although essential to economic growth, can lead to deeper economic and social change affecting the values and drivers behind the concept of sustainable development. Thus, new and more sustainable life standards, new consumption patterns and firms' commitment to the natural environment do not appear independently from knowledge, but on the contrary, are strongly connected to it.

In parallel with the consolidation of RBV as a prominent theoretical approach for strategic management scholars, Hart (1995) argued the necessity of including natural environmental (and social) issues in the heart of RBT proposing a natural one. According to Hart's vision, the RBV incurs a major omission by ignoring the interaction between the firm and the natural environment. Although this omission could be considered as acceptable in the past, nowadays it is evident that the natural environment is a factor that may constrain the achievement of competitive advantage. In other words, strategy and competitive advantage in the coming years shall rest on the capabilities that facilitate environmentally sustainable activity. Therefore, the RBV needs to address important current economic and social phenomena. Among these new concerns, we can highlight the business at the Bottom of Pyramid (BoP) which necessarily implies new and innovative business models or the inclusion of sustainability concerns in firms' decision-making process in order to maintain competitive positions and avoid environmental and social reputational risks.

Finally, but not less important, a theory – such as RBV – needs an empirical test in order to be considered a real theory. Nevertheless, the literature review offers a great deficit in that sense. Among possible causes, the intangible character of the most strategic relevant resources and capabilities (dynamic capabilities, innovation capability, corporate reputation, organizational culture, organizational learning, absorptive capacity, knowledge, human capital, etc.) impedes their effective measurement and assessment. In that sense, since the 1990s (Bontis, 1998; Subramanian and Youndt, 2005), different practitioners and academics tried to overcome those difficulties, developing intellectual capital measurement models, which can be labelled as an 'Intellectual Capital-Based View' (ICBV) (Reed, Lubaking and Srinivasan, 2006). They conceive ICBV as a middle-range theory that tries to overcome RBV's methodological problems. ICBV focuses on measurement of intangible or intellectual resources or knowledge assets, organizing them in three main categories: human capital, which includes individual's knowledge, abilities, skills and behaviours; social capital, which embodies the knowledge inherent in social and network ties among individuals, inside and outside the organization; and organizational capital, which includes the institutionalized collective knowledge embedded in databases, processes, routines and information technologies (Martín-de Castro et al., 2011).

The ICBV and the natural environment have been, until recently, different fields walking on their own different directions and only few scholars have addressed the connection between these two realities. Although still in a nascent stage, the research about these intellectual capital-natural environment connections constitutes a promising avenue for future development. In this sense, the work of Chen (2008) is one of the more remarkable efforts to show the importance of green intellectual capital to competitive advantage achievement. Drawing on the intellectual capital classification of Johnson (1999) and Bontis (1999), the author differentiates green intellectual capital into green human capital, green structural capital and green relational capital, arguing that these three types of green intellectual capital have positive effects on the competitive advantage of firms.

Thus, green human capital understood that employees' skills, stocks of knowledge, attitude or commitments about environmental issues play a great role in the development of environmental innovations or in the environmental management in general. They are particularly important in the pollution prevention strategies noted by the NRBV theory, where continuous improvement and employees' attitudes towards sustainable development are crucial for its adequate development.

In the same way, green structural capital in terms of organizational culture, company images, knowledge management systems, databases or technology systems containing very valuable information in environmental terms (e.g. pollution records, water and energy consumption) are part of the organizational endowment that support both the environmental management of the company and the development of environmental strategies.

In turn, green relational capital also plays a critical role in the environmental management of the company. As one of the main goals of the NRBV is the spread of environmental concern all over the value chain, the relationships with customers, suppliers and institutions are of the utmost importance. The great value of this kind of relationships can be demonstrated through the product stewardship strategies in which the design of the product and also the collaboration of the stakeholders through its value chain can be a source of competitive advantage.

In a wide sense, all these ramifications, 'spin-offs' and/or theoretical developments reflect the current development and strong power of explanation that RBT has for both practitioners and management academics. And one of the most promising developments, as this research work highlights as its main thesis, is the role of natural and environmental sustainability of a firm's business activities, as the NRBV proposes.

1.3.2 Steps to corporate sustainability

According to the NRBV there are three strategic capabilities[3] called pollution prevention, product stewardship and sustainable development, each one of them targeting a particular purpose, articulated around different key resources and with different sources of competitive advantage (Hart and Dowell, 2010) (Figure 1.2).

Pollution Prevention seeks to reduce waste and emissions generated during the production process rather than controlling them once they have occurred. Usually, it is associated with lower costs and efficiency improvements in the production process as it enables the reduction of required inputs, streamlining processes and reducing compliance cost and environmental risk arising from operations. As stated by Hart (1995) pollution prevention has similarities in some respects to Total Quality Management, as both require the intensive involvement of employees and continuous improvement in reducing emissions, rather than expensive pollution control technology.

Prevention emphasizes efficiency and productivity improvements since less waste is generated, giving a better use to the production process inputs and reducing production cycles by means of simplifying or removing unnecessary steps.

Strategic capability	Environmental driving force	Key resource
Pollution prevention	Minimize emissions effuents & waste	Continuous improvement
Product stewardship	Minimize life cost of products	Stakeholder integration
Sustainable development	Minimize environmental burden of firm grouth and development	Shared vision

Figure 1.2 NRBV: Conceptual framework
Source: Hart (1995: 992).

From a resource-based point of view, it is argued that pollution prevention strategies generate more benefits to the firm if they are accompanied by strong innovation capabilities (King and Lenox, 2002). Therefore, the prevention commitment of the firms will not generate positive economic returns rents by itself, but in combination with innovation capabilities may lead to positive results (Christmann, 2000).

In turn, Product Stewardship broadens the pollution prevention scope as it takes into consideration the entire product value chain. So, with the main aim of reducing product environmental cost through the life cycle, product designers should minimize the use of non-renewable materials, avoid the use of toxic components, reduce the environmental impact during the use and configure the products to be easily reused or recycled at the end of its life.

Using this life cycle approach 'from the cradle to the grave' seeks to involve stakeholders in the entire product development cycle (Sharma and Vredenburg, 1998), leading to potential competitive advantages derived from a privileged position that allows, for example, preferential access to raw materials or to establish certain standards in the industry.[4]

Nevertheless, research in this area is still in its early stages, being a newly emerging and promising field (Hart and Dowell, 2010). Therefore, most contributions in this area have used case studies, being the presence of more comprehensive works almost symbolic.

Sustainable development, however, differs from pollution prevention and product stewardship in two major respects. First, it emphasizes the indefinite nature of environmental measures, and second, it is not restricted solely to environmental issues but also takes into consideration the impact on the economic activity of developed countries over the less developed, causing degradation and poverty (Hart and Christensen, 2002). Thus, a sustainable development-oriented strategy must recognize this reality and act both to reduce the environmental impact of economic activity and to improve the quality of life of those disadvantaged groups who are affected by more developed countries' business practices.

Based on his previous contribution, Hart (1997) drew a clear distinction between the above-mentioned environmental alternatives, depending on current or future orientation and on the pursued degree of change (Figure 1.3). From this perspective, both pollution

20 Environmental Innovation and Firm Performance

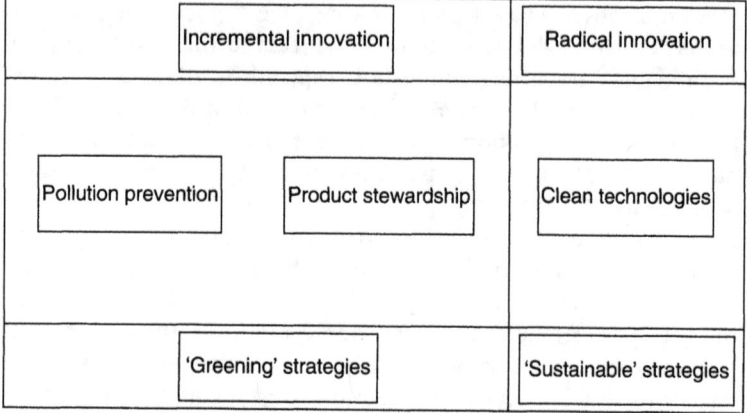

Figure 1.3 Environmental innovation and sustainable strategies
Source: Own elaboration from Hart (1997: 74).

prevention and product stewardship are oriented to current products and processes, giving rise to the so-called *greening strategies*, while sustainable development and clean production are future oriented (future markets and technologies), resulting in the *beyond greening strategies*.

Thus, while *greening strategies* lead to incremental changes, sustainable strategies seek radical (clean technologies) or disruptive changes (Christensen et al., 2006) aimed at integrating in the market (as active consumers) to the most disadvantaged groups, also known as the Bottom of the Pyramid. Using a different but economically feasible business logic, disruptive innovations offers products and services that aren't as good as those in mainstreams markets (Hart and Christensen, 2002). In fact, these disruptive products and services are oriented to non-traditional customers, where industry leaders are not willing to be and find in the groups at the base or bottom of the pyramid the perfect place to be developed.

Finally, the scope of the NRBV would not be fully represented if we did not also consider the Dynamic Capabilities approach, which can be viewed as a RBV evolution that moves parallel to the sustainable strategies recently discussed.

The concept of dynamic capability acquired great relevance in the management literature with the work of Teece et al.

(1997). From this perspective, the Dynamic Capability approach is conceived as an extension of the RBV (Barney, 1986, 1991), which tries to fill the already mentioned 'static gap' of the RBV (Priem and Butler, 2001).

As argued by Barreto (2010), the dynamic capabilities revolve around three main assumptions:

- The idea of change: The constant reference to the idea of change and resource configuration is common among dynamic capabilities scholars (Teece et al., 1997; Eisenhardt and Martin, 2000; Helfat et al., 2007; Barreto, 2010). Besides the time factor, Eisenhardt and Martin (2000) argue that organizational change must be carried out before the competitors.
- Opportunities and threats identification: The dynamic capabilities approach, through its monitoring capabilities, allows us to identify the opportunities and threats emerging from the changes in the environment (Barreto, 2010).
- Systematic character: The dynamic capabilities allow the resolution of problems in a systematic, structured, persistent and nonrandomized manner (Barreto, 2010).

Taking as the starting point the work of Teece et al. (1997) and their concern about how businesses can renew their competitive sources in rapidly changing environments, Aragón-Correa and Sharma (2003) emphasized the importance of clean technology and Bottom of Pyramid oriented strategies stressing the fact that, from their point of view, the Dynamic Capabilities approach fits perfectly with this kind of strategy as the environments in which they take place are ambiguous and highly complex. In this sense, the degradation of the ecosystems and the resource depletion create discontinuities that threaten firms' current resources and capabilities endowments. Besides this, the challenges associated with the groups at the Bottom of Pyramid (characterized by the lack of institutions, poor infrastructures and low literacy levels) require that firms develop entirely new capabilities in order to successfully create the appropriate business models to serve these groups (Hart, 2007, in Hart and Dowell, 2010).

Thus, as clean technologies involve the necessary skills to deal with fields of knowledge that are uncertain and constantly evolving, firms must protect and also draw on radical innovations that may even eliminate some of its core businesses. Following the same argumentation, as firms are moving into markets where there are completely different business logics (as it happens in the Bottom of Pyramid markets), they will have to address new problems and also new solutions (Hart, 1995; Aragón-Correa and Sharma, 2003).

As shown, the NRBV scope is quite broad, addressing many aspects that range from firm-stakeholders' interaction to the adaptation to changing environments through green or sustainable strategies. The joint treatment of these issues goes far beyond the scope of this book, which will be specifically detailed in the next section.

1.3.3 Defining the scope

One of the main research lines in the environmental field revolves around the question of whether it is economically feasible for businesses to incorporate environmental criteria into their activities (Hart and Ahuja, 1996). To address issues arising from this question, the NRBV provides the right theoretical foundations from which firms can build bridges between environmental and financial performance, so that the link between environmental strategy and competitive advantage depends on the type of environmental improvement that is taken into consideration. Therefore, the effects derived from this environmental-financial link will be different as we try to implement the above-mentioned greening or beyond greening strategies.

As previously noted, environmental resources and capabilities play a major role in environmental strategy. According to Russo and Fouts (1997) environmental active (prevention) and environmental reactive (control based) firm's resources and capabilities endowments are quite different. Specifically, the empirical studies conducted to date argue that innovation capabilities, including those related to continuous improvement, enable the firm to stay one step ahead of competitors (Sharma and Vredenburg, 1998) and make more likely the achievement of economic returns derived from the pollution prevention strategy (King and Lenox, 2002).

In the same line, the main aim of this book is to provide further evidence of such empirical work, emphasizing, as a novel contribution in the field, the role played by some of the least-researched facets that make up the product stewardship strategy, that is, environmental product innovation and green image. To do that, first we analyse the environmental product innovation efforts carried out by the firms, and second, the major role played by firm's green image on the task of involving the stakeholders in the development of the entire value chain of the product.

Following Morrison and Siegel (1997, in King and Lenox, 2002), for which pollution prevention efforts can be a proxy for environmental capabilities, in this book we will extend the scope of that claim up to certain aspects of product stewardship. To reach that objective, we will use the environmental innovation as the connecting link.

2
The Nature of Environmental Product Innovation and Green Image

2.1 Introduction

From the second half of the 1990s concern for the relationship between firms and the natural environment has been growing in the academic literature (Hart and Dowell, 2010). Although industrial development over the last two years has brought prosperity and wealth, it has caused environmental degradation unintentionally (Shrivastava, 1995). Industrial activity has increased to the point that there have been irreversible effects on our global environment, including impact on climate, biodiversity and ecosystems. Global population growth has resulted in mass migration from rural areas to cities, increasing social and economic inequalities and social degradation (Hart and Milstein, 2003).

Therefore, it is necessary to change current growth patterns and respond as soon as possible to the challenge of sustainable development. In other words, our society must turn to what the Brundtland Commission defined as the kind of development that meets present needs without compromising future generations (Sharma and Vredenburg, 1998).

Thus firms, as one of the main society actors cannot stay away from this important challenge. They must assume their responsibility, putting in practice a wide range of measures aimed to reduce, on the one hand, polluting emissions and material consumption levels, and on the other hand, to develop new and more efficient green technologies that can, at the same time, create value to customers and shareholders.

Taking these ideas as our main reference point, in the next section we will emphasize the role of environmental innovation as one of the firm's key contributions to sustainable development. Through an extensive literature review, first we will explore how scholars have defined the term 'innovation', trying to underline the existence of three different approaches (economic, social and socio-economic) to the concept. Consequently, stressing the socio-economic character of environmental innovations, we will conduct an in-depth analysis of environmental product innovation and green corporate image as two of the main elements of Hart's product stewardship strategy, which is at the centre of the Natural Resource-Based View (NRBV).

2.2 Innovation: Approaches to the term

From the beginning of the last century, several orientations have been used to define the concept of innovation (Table 2.1). As a consequence, there is a lack of consensus and the contributions are quite fragmented. Faced with this reality, our intention is to classify and group these widely spread contributions.

Table 2.1 Innovation definitions

Author	Definition of Innovation
Schumpeter (1912)	Implementation of new combinations of materials and forces, which may include the introduction of a new and untested product or production method, developing a new market, the conquest of a new source of supply, and creating a new organization in any industry
Thompson (1965)	Generation, acceptance and implementation of new ideas, processes, products and services
Knight (1967)	Adoption of a new and significant change by an organization
Nelson (1968)	Process by which new products and techniques are introduced into the economic system
Zaltman et al. (1973)	Any idea, practice or material appliance perceived as being new to the organizational unit that adopts it
Rowe and Boise (1974)	Successful use of products or processes that are new to the organization and that achieve results according to decisions made within it

Table 2.1 (Continued)

Author	Definition of Innovation
Gee (1981)	Process by which, from an idea, an invention or the identification of a need, a product, technology or service is developed that is accepted commercially
Damanpour and Evan (1984)	Deployment of an idea related to a device, system, process, policy, programme or service that is new to the organization at the time of adoption
Tushman and Nadler (1986)	Creation of any product, service or process that is new to a business unit
Van de Ven (1986)	A new idea, which may be a combination of old ideas, a project that challenges the present situation, a formula or a unique approach that is perceived as being new by the individuals involved
Deward and Dutton (1986)	Idea, practice or device received by the relevant unit of adoption as new material
Damanpour (1991)	Generation, development and implementation of new ideas or behaviours
OECD (1992)	Transforming an idea into a successful market product, new or improved, or a business process in industry and trade or a method of social service
Nonaka (1994)	Process in which the organization defines and creates problems, and develops new knowledge to solve them
European Commission (1995)	Means to produce, assimilate and operate a successful innovation in the economic and social fields that brings new solutions to problems and thus meets the needs of both individuals and society
Morcillo (1995)	Seeing what everyone sees, reading what everyone reads, hearing what everyone hears, to innovate is to make what no one has yet imagined
Noria and Gulati (1996)	Any policy, structure, method or process, or any product or market opportunity seen by the manager of a unit as innovative
Freeman and Soete (1997)	Attempts to commercialize an invention (an invention being the discovery of new methods or materials, namely, the discovery of new knowledge)
Galunic and Rodan (1998)	Re-conceptualization of an existing system to use resources, which is developed as a new way of generating income and potential
Damanpour and Gopalakrishnan (1998)	Adoption of a new idea or behaviour in an organization

Williams (1999)	Implementation of discoveries and inventions, as well as the process by which new results are born, whether products, systems or processes
Camelo et al. (2000)	Creation or acquisition of an idea or knowledge, and its introduction into the organization, which may materialize in the form of a new product, process or method
Nieto (2001)	First application of an invention that is produced when the first commercial transaction with the new products processes or services derived from it
Johnson et al. (2002)	Successful introduction of an invention in production or the market
Chen et al. (2004)	Introduction of a new combination of essential factors of production in the production system. It involves new products, new technologies, new markets, new materials and new combinations
Edvinsson et al. (2004)	Re-using existing knowledge and perspectives in combination with new knowledge, as an invention, and then commercializing and capitalizing on it
Carson et al. (2004)	Creation of new knowledge that has potential for practical application in the development of new products or processes
Subramanian and Youndt (2005)	Identifying and exploiting opportunities to create new products, services or work practices
OECD (2006)	Introduction of a new or significantly improved product (or service) process, marketing method or methods in organizational practices within the company, in the workplace or in foreign affairs
Grant (2006: 408)	Initial marketing of an invention to produce and market a new product or service or use a new method of production
Egbetokun et al. (2009)	A process by which firms master and implement the design and production of goods and services that are new to them, irrespective of whether they are new to their competitors, their countries or the world
Tödtling et al. (2009)	The result of an interactive process of knowledge generation, diffusion and application

Source: Adapted from Delgado (2009).

According to our theoretical framework, taking into account the constraints imposed by the natural environment is not only good for the business but also for society in general. This is why we have chosen a classification of innovations that reflects this double economic and social reality. Therefore, we can differentiate the contributions in three separate blocks depending on the economic, social or socio-economic orientation of the term.

Most of the contributions above highlight the economic aspects of the term (Table 2.2). Thus, this wide group of definitions relates innovation to firms, marketing, products and processes and implicitly relies on economic considerations. In this line we find contributions belonging to different time periods like Schumpeter (1912), Thompson (1965), Nelson (1968), Noria and Gulati (1996), Camelo et al. (2000), Nieto (2001) or Carson et al. (2004), among others. There exist also, within this approach, some contributions that explicitly mention the generation of rents to define the concept (Galunic and Rodan, 1998).

Furthermore, other groups of scholars define innovation from a social perspective (Zaltman et al., 1973; Deward and Dutton, 1986; Van de Ven, 1986; Damanpour, 1991; Morcillo, 1995; Tödtling et al.,

Table 2.2 Innovation orientations

Innovation Orientation	Authors
Economic	Schumpeter (1912), Thompson (1965), Knight (1967), Nelson (1968), Rowe and Boise (1974), Gee (1981), Damanpour and Evan (1984), Tushman and Nadler (1986), Deward and Dutton (1986), Nonaka (1994), Noria and Gulati (1996), Freeman and Soete (1997), Galunic and Rodan (1998), Damanpour and Gopalakrishnan (1998), Williams (1999), Camelo et al. (2000), Nieto (2001), Johnson et al. (2002), Chen et al. (2004), Edvinsson et al. (2004), Carson et al. (2004), Subramanian and Youndt (2005), OECD (2006), Grant (2006), Egbetokun et al. (2009)
Social	Zaltman et al. (1973), Van de Ven (1986), Damanpour (1991), Morcillo (1995), Tödtling et al. (2009)
Socio-economic	OECD (1997), EC (1995)

Source: Own elaboration.

2009). As a common feature to all of them we found the emphasis was on novelty, stressing the role of novelty perceptions. The contributions within this group range from those that mention new practices that are perceived as innovative (Zaltman et al., 1973) to new ideas and behaviours (Damanpour, 1991), interactions with the environment (Morcillo, 2005) and new knowledge generation (Tödtling et al., 2009).

Finally, in a third group we place the definitions which emphasize the socio-economic orientation of the term. In this sense, the definition given by OECD (1997), besides referring to new processes and products also mention new ideas that lead to new social service methods. Also, the EC (1995) mentions the successful exploitation of new developments in the economic and social spheres, highlighting the response to people's and society's needs. This latter orientation fits perfectly with the purpose of our book as it covers the entire scope of our theoretical framework and is in line with NRBV strategies. Accordingly, to the exploration of this socio-economic orientation we will devote the next section.

2.3 Socio-economic innovations

Among the first and most important contributions that make explicit reference to the socio-economic objectives of the innovations we can highlight those coming from the OECD (1992) and the EC (1995). According to their point of view, innovations can transform ideas into marketable products or business processes that lead, in turn, to new methods of social service or successful exploitation of new developments.

In the same line, some relevant contributions coming from the Social Innovation field emphasize as well the socio-economic character of innovations (Table 2.3). In this sense, some scholars categorize like innovations concepts such as lifestyle changes or new customers' behaviour (Scherhorn et al., 1997, in Rennings, 2000). In the same vein, Duchin (1999) relates social innovations with new technologies and new dynamic lifestyles, Hamalainen and Heiscala (2007) argue that social innovations lead to better social and economic results and institutions like the Centre for Social Innovation (2008), following previous authors, call 'social' those innovations aimed at solving the current economic, social and cultural challenges.

Table 2.3 Socio-economic innovations

Author	Definition
Scherhorn et al. (1997)	Social Innovations as the changes in lifestyles or behaviour of consumers
Duchin (1999)	Social Innovation as new technologies and new lifestyle dynamics
Forum on Social Innovation (OCED) (2000)	Innovations whose main aim is to satisfy unmet social needs
Hamalainen and Heiscala (2007)	Social innovation as the changes in social structures that enhance the social and economic performance and the collective power of resources
Centre for Social Innovation (2008)	Those new ideas that address current economic, social, cultural and environmental challenges to benefit the planet and the people who inhabit it
Yunus et al. (2010)	Building social innovation models require two additional specificities: (i) favouring social profit-oriented shareholders and (ii) clearly specifying the social profit objective
Dawson et al. (2010)	Social innovations are triggered by an interest in improving the well-being of people in society. Its aim is to improve the welfare of groups and communities, as such they may; seek to further the social conditions of work; hope to provide socially useful solutions to ongoing community problems; or provide improvements in well-being for remote or socially isolated communities

Source: Own elaboration.

More recent contributions also revolve around the same idea. Yunus, Moingeon and Lehman-Ortega (2010) argue that, to be successful, social innovations must define clearly its social objective and also must look for the satisfaction of its more socially oriented shareholders. According to Dawson, Daniel and Farmer (2010), social innovations are aimed at improving the welfare of society, groups and communities and working conditions, which shows the alignment of these authors with The Forum on Social Innovation and its Local Economic and Employment Development (LEED) programme, whose primary objective is the satisfaction of social needs not covered by the market and the creation of successful insertion ways for people outside the production systems.

As shown, socio-economic innovations are perfectly compatible with business activities and rent generation. As argued by Porter and Kramer (2006), the fact of adding a social dimension to the value proposition of the firm offers new possibilities in the firm's competitive landscape.

Therefore, the NRBV is the perfect framework for the development of these socio-economic innovations. Thus, innovations aimed at the Bottom of the Pyramid (Hart and Milstein, 2003), Embedded Innovations (Simanis and Hart, 2009) and Clean Technologies and pollution prevention strategies, among others, definitely belong to this category.

Bottom of Pyramid innovations aim to achieve socio-economic objectives. With the creation of inclusive businesses two main goals are met. On the one hand, the social integration of the most disadvantaged collectives, and on the other hand their incorporation to the market as active consumers. This approach is instrumental in nature (Garriga and Melé, 2004), and claims that firms that make the leap to the Bottom of Pyramid will not only do the right thing but also will find a huge market full of opportunities (Hart and Christensen, 2002). Firms developing this strategic option will gain access to these markets by offering regular products with basic features adapted to the needs of these non-traditional customers. The economic argument behind these socio-economic innovations is that, once established in these new markets with many potential customers, firms would develop new business models and capabilities (Mutis and Ricart, 2008) also accessing higher income segments and improving their margins.

Also, with the same socio-economic approach, we find the Embedded Innovation (Simanis and Hart, 2009), whose basis lies in the creation of new communities characterized by a very sharp sense of belonging. Therefore, the value of these new and innovative business models lies in the very concept of community. According to this approach, firms will have to dive into communities, with the main aim of creating a joint business concept characterized by mutual learning and cooperation between the firm and the community members. One key feature of these innovative business models is that they are not only applicable to the markets at the Bottom of the Pyramid but also to the more conventional markets through the implementation of Catalytic Innovations (Christensen et al., 2006), whose main aim is to provide social services to the most neglected segments (not attractive to large companies) of population within developed economies. Behind this catalytic approach is the fact that in some 'developed' countries there exist a misdirected investment effort in social needs. Looking for sustainable and what they call system-changing solutions, the authors posit the catalytic innovations drawing on Christiansen's disruptive-innovation model. Following this model, the catalytic innovations are a subset of disruptive innovations, with the main aim of promoting social change by offering simpler and good enough services and solutions to underserved customers. Although the authors focus their interest on certain specific solutions like health care, education or microlending, the model can be adapted to other social sectors where the dominant provider is moving away towards more profitable segments of the market.

Among the most important socio-economic innovations that link environmental change and the generation of business opportunities we can find clean technologies, pollution prevention strategies and innovations aimed at Sustainable Development (Nidumolu et al., 2009), which emphasize the idea of change in the firm's relationships with both its natural and social environment. Thus, those companies that understand sooner this need for change will be best placed to drive tomorrow's economy.

Therefore, environmental proactive strategies like clean technologies and pollution prevention represent an innovative way to face the challenges and constraints resulting from the future scarcity of natural resources. Innovation, continuous improvement and cost minimization can lead firms to new competitive advantage sources, improving productivity and efficiency and creating new and highly competitive capabilities (Hart, 1995; Hart and Ahuja, 1996; Sharma and Vredenburg, 1998; Nidumolu et al., 2009). Therefore, environmental innovations, as we will see in the next section, fall into this category.

2.3.1 Environmental innovation. Definition, determinants and types

Among the innovations of socio-economic character, we want to highlight the role played by environmental innovations. In general terms, it is often argued that these innovations are conditioned both by the technological possibilities of the firm and its ability to appropriate the benefits of innovative activity (Horbach, 2008). One of its main features is the so-called 'double externality', since beside the generation of the positive externalities that are common to all innovations, environmental innovations benefit the society as a whole (Rennings, 2000).

It is, therefore, the social benefit produced which gives the socio-economic character to these innovations. Thus, together with environmental technologies, changes in lifestyles and consumer habits are essential to reduce the impact of human activity on the planet.

Environmental innovation incorporates ethical arguments to products, processes and organizational modes of the company. This statement, however, lacks specificity. Therefore, it is necessary to specify in more detail its nature, its determinants and the different types of environmental innovations in order to understand the strategic options in the environmental field.

Some academics suggest that the term 'environmental innovation' has different natures (Rennings, 2000). Thus, its nature can be technological, organizational, institutional or social. Technological nature can be seen in environmental technologies for the prevention of environmental pollution or for the control of the same, commonly called 'end of pipe'. In relation to its organizational nature it includes environmental management tools such as International Organization for Standardization (ISO) 14001 or the Eco-Management and Audit Scheme (EMAS). Institutional and social nature can be seen in the environmental institutions such as the intergovernmental panel on climate change and in changes in styles and dynamics of life and consumption, respectively.

Due to their broad scope, environmental innovations have been addressed from different angles (Table 2.4); there exist many contributions ranging from very narrow conceptions of the term that focus on 'green' products or processes (Chen et al., 2006) to very broad approaches that go beyond the limits of the firm involving changes in socio-cultural norms (OECD, 2009).

As shown above, most environmental innovation definitions refer to products, processes or management practices aimed to reduce the environmental impact (Kemp and Arundel, 1998; Rennings and Zwick, 2002; European Commission, 2008; Kemp and Pearson, 2008).

Table 2.4 Environmental innovation definitions

Author	Definition
Fussler and James (1996)	The process of developing new products, processes or services which provide customer and business value but significantly decrease environmental impact
Hemmelskamp (1997)	Innovations aimed to reduce the negative environmental impacts caused by production methods
Klemmer et al. (1999)	All measures of relevant actors (firms, politicians, unions, associations, churches, private households) which develop new ideas, behaviour, products and processes, apply or introduce them and which contribute to a reduction of environmental burdens or to ecologically specified sustainability targets
Vinnova (2001)	Innovation that serves to prevent or reduce anthropogenic burdens on the environment, clean up damage already caused or diagnose and monitor environmental problems
Andersen (2002)	Innovation which is able to attract green rents on the market
Kemp and Arundel (1998), Kemp (2001), Rennings and Zwick (2003)	New and modified processes, equipment, products, techniques and management systems that avoid or reduce harmful environmental impacts
Huber (2004)	Technological environmental innovations (TEIs) may help to reduce the quantities of resources and sinks used, be they measured as specific environmental intensity per unit of output, or as average consumption per capita or even in absolute volumes
European Commission (2004)	Environmental technologies include all those whose use is less environmentally harmful than relevant alternatives
Little (2005)	'Sustainability-driven' innovation is 'the creation of new market space, products and services or processes driven by social, environmental or sustainability issues'
Chen et al. (2006: 333)	Hardware or software innovation that is related to green products or processes

Europe INNOVA (2006)	Europe INNOVA is an initiative of the European Commission (Enterprise and Industry). 'Eco-innovation is the creation of novel and competitively priced goods, processes, systems, services, and procedures designed to satisfy human needs and provide a better quality of life for all, with a life-cycle minimal use of natural resources (materials including energy, and surface area) per unit output, and a minimal release of toxic substances'
European Commission (2007)	'Eco-innovation is any form of innovation aiming at significant and demonstrable progress towards the goal of sustainable development, through reducing impacts on the environment or achieving a more efficient and responsible use of natural resources, including energy'
Kemp and Pearson (2008)	'Eco-innovation is the production, assimilation or exploitation of a product, production process, service or management or business method that is novel to the organization (developing or adopting it) and which results, throughout its life cycle, in a reduction of environmental risk, pollution and other negative impacts of resources use (including energy use) compared to relevant alternatives'
European Commission (2008)	Eco-innovation is 'the production, assimilation or exploitation of a novelty in products, production processes, services or in management and business methods, which aims, throughout its lifecycle, to prevent or substantially reduce environmental risk, pollution and other negative impacts of resource use (including energy)'
Oltra and Saint Jean (2009)	In a broad sense, environmental innovations can be defined as innovations that consist of new or modified processes, practices, systems and products which benefit the environment and so contribute to environmental sustainability
OECD (2009)	Eco-innovation is generally the same as other types of innovation but with two important distinctions: 1) Eco-innovation represents innovation that results in a reduction of environmental impact, whether such an effect is intended or not; 2) The scope of eco-innovation may go beyond the conventional organizational boundaries of the innovating organization and involve broader social arrangements that trigger changes in existing socio-cultural norms and institutional structures

Source: Adapted from Carrillo Hermosilla et al. (2009).

As mentioned by Hellström (2007) these eco-innovations can also be differentiated in a number of ways. For example, eco-innovations can be incremental, radical or disruptive depending on the newness of the offering and on how systemic they are. In this sense, as process innovations are usually associated with efficiency gains, it seems that they are more related to incremental rather to radical changes. Furthermore, it is also assumed that incremental innovations will have decreasing marginal returns, a circumstance that makes these innovations highly dependent on radical innovations which will regularly push the technological system up. In turn, disruptive innovations are the ones that make current systems obsolete. Kemp and Foxon (2007) give good examples of all these eco-innovations. For example, steps reduction in the production process can be considered as incremental while fuel injection is a radical but not disruptive innovation because it is used within an existing system. However, in-wheel electric propulsion is a disruptive innovation as it makes the internal engine and all the activities around it obsolete. Nevertheless, it is also true that the line between all these eco-innovations is very thin and sometimes its labelling is a matter of perspective. One good example of incremental innovation is the Responsible Care programme in the chemical industry (Olcese, 2008), through which most firms in this industry, trying to prevent the return of the shocking and tragic incidents that happened in Union Carbide (Bophal) or Hoffman-La Roche (Seveso), implemented procedures and management systems to help companies develop their pollution prevention strategies, reduce the environmental impact of the activities and cooperate with the communities in which they operate. As an additional example of radical changes within the same sector we can mention the initiatives carried out by leading companies like DuPont or Novo Nordisk whose commitment to the future is based on the development of renewable materials. In this sense the development of new biomass and biological enzymes based plastics represent a radical change in the chemical industry.

Furthermore, incremental and radical eco-innovations can also be component based or architectural based. As Hellström (2007) shows, measures aimed to increase eco-efficiency in existing processes (e.g. introduction of replacement materials in production) are examples of incremental-component eco-innovations. Incremental-architectural are, in turn, based in system changes in order to improve efficiency as, for example, new ways of building car washers that reuse water or new water-recovering methods that enable houses to accumulate rainwater so it can be used to cover families' non-essential water necessities.

Radical-component eco-innovations are those that replace one critical component with a completely new solution. For example, the new and more environmentally benign processes for heating up objects on a production line can be considered as radical-component eco-innovations. Finally, radical-architectural eco-innovations address environmental problems in a radically new way that restructures the architecture of an old process. One example of this latter category can be inclusion of new methods for water purification with magnetic induction.[1]

According to Chen et al. (2006), green innovations are hardware or software innovations related to green products or processes, including the innovation in technologies that are involved in energy-saving, pollution prevention, waste recycling, green product design or corporate environmental management. Kemp et al. (2001, in Horbach, 2008) argues that environmental innovations consist of new or modified processes, techniques, systems and products to avoid or reduce environmental damage, and according to Rennings (2000), environmental innovations can be defined as the measures of relevant actors consisting in the development, application or introduction of new ideas, behaviours, products and processes that contribute to a reduction of environmental burdens or to ecologically specified sustainability targets.

While all of these contributions are very valuable, we argue instead that the social aspect of environmental innovations (e.g. proactive environmental behaviours that happen outside the firm) must be taken into consideration.

Therefore, we argue that the definitions given by Klemmer et al. (1999, in Carrillo-Hermosilla et al., 2010) and OECD (2009) are complementary in nature and, placed together or jointly analysed, as they refer not only to technological innovations but also to changes in institutions and norms, explain properly the global scope of the term.

Thus, taking into account all the above, environmental innovations are those measures or activities consisting of the development, application or introduction of new ideas, behaviours, products, processes, procedures and organizational systems that contribute to the reduction of the environmental impact making possible changes in socio-cultural norms and institutional structures.

Regarding the determinants of environmental innovation, the literature suggests that these are supply factors, demand factors and institutional and political influences (Horbach, 2008). From the supply side, the environmental innovations (as many innovations) are conditioned by the available technological possibilities of the firm and by the return appropriation of the innovation activities. In this regard, the double externality problem must be highlighted. Environmental innovations, besides the positive externalities from spillovers which are common to all innovations, are characterized by the fact that while the whole society benefits from a technical environmental innovation, the costs have to be borne by a single firm (Rennings, 2000).

From the demand side, both the potential market demand and the social awareness can determine the posture of the firm regarding the Environmental Innovations. Thus, through the Environmental Innovation, firms may have access to those segments of the market willing to pay a premium for green products (Miles et al., 1997) and also create a green reputation (Chen, 2008). Furthermore, companies can leverage their reputation for environmental innovation to gain preferential access to new and lucrative businesses like waste management, recycling services and environmental impact analysis among others (Nidumolu et al., 2009).

In relation to the institutional and political influences, the role of the environmental regulation should be noted. First, environmental regulation may force firms to realize economically benign Environmental Innovation, and second, firms may find early movers advantages from adapting to regulation before than their rivals (Porter and Van der Linde, 1995; Horbach, 2008).

In addition to the character and determinants of environmental innovations, we must refer to the environmental innovation types in order to show a complete picture of the topic. Thus, following the OECD (1997) guidelines, we can distinguish between technical and organizational innovations. Thus, technical environmental innovations are specific kinds of innovations that consist of new or modified products and processes to avoid or reduce the environmental burden, while environmental organizational innovations include

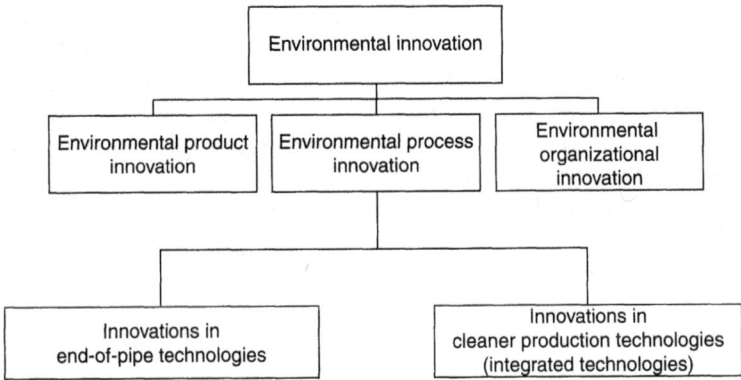

Figure 2.1 Environmental innovation types
Source: Rennings et al. (2006: 47).

the re-organization of processes and responsibilities within the firm with the objective to reduce environmental impacts (Rennings et al., 2006) (Figure 2.1).

Among technical environmental innovations we can find environmental process innovations and environmental product innovations. As process innovations we can include those aimed at reducing energy consumption during the production process or those that convert waste into new ways of creating value, both through its reuse within the firm or outside, selling the waste in those businesses where they could be useful (Porter and Van der Linde, 1995). Among environmental process innovations we can include the reductions in air or water emissions, improvements in resource and energy efficiency, reductions in water consumption and switching fossils fuels to bioenergy (Kivimaa and Kautto, 2010).

In general terms, environmental process innovations can be subdivided into innovations in end-of-pipe technologies and innovations in integrated technologies (also called cleaner production technologies). While end-of-pipe technologies are oriented to comply with environmental regulation (waste disposal, water protection, noise abatement or air quality control technologies), cleaner production technologies emphasize continuous improvement and cost minimization. Examples of cleaner production technologies are the recirculation of materials, the use of environmentally friendly materials and the modification of the combustion chamber design. Some authors, adding more detail to the technological environmental innovations typology, distinguish between end-of-pipe integrated (preventive) and end-of-pipe non-integrated (control) depending on whether these technologies are integrated in the production process or not, Hartie (1990, in Hemmelskamp, 1997).

Among technological innovations, environmental process innovations have been, in comparison, much more analysed than environmental product innovations. Product design innovations like those responding to the concept of 'design for disassembly' (Shrivastava, 1995) (which is based on creating products that are designed for easy recovery, dismantling and recycling, thus extending the life of each of the components), improvements in the durability of the products, raw materials reductions, selection of environmentally less harmful raw materials and removal of hazardous substances (Kivimaa and Kautto, 2010) have so far received much less attention in the literature, and it is one of our main objectives in this book to shed some light on this matter. Accordingly, in the next sections we will emphasize its importance both to theory and practice.

Environmental organizational innovations, in turn, as we will see later on, can be supporting factors for technical environmental innovations. Among them, we can mention as one of the most prominent initiatives the utilization of environmental management systems (EMS) like EMAS or ISO 14001 and the 'green teams' which are composed of members of the organization from various departments and levels of responsibility, whose job is to advise the company on the impact of their activities on the environment. This advisory work covers all areas of business activities and includes the development of programmes for waste management, energy and resource conservation or renewable energy source exploration (Shrivastava and Hart, 1995).

2.3.2 Environmental process innovations

Recent studies have shown that environmental innovations, in the most polluting industries, were more focused on processes than on products. This evidence is partly explained because regulation and consumer pressures are mainly focused on process and not on product (Kivima and Kautto, 2010).

Environmental process innovations, pollution prevention oriented, aim to reduce the environmental impact of firms' operations. Focused mainly on process changes and modifications 'during the process' are cost reduction and efficiency oriented (Rennings et al., 2006; Frondel et al., 2007). Some examples of this kind of innovations can be found in the academic literature when scholars refer to several practices like preventive equipment maintenance, material optimization and stage reduction during the production process (Shrivastava, 1995; Christmann, 2000; González-Benito and González-Benito, 2008).

It is also argued that this kind of innovation promotes continuous improvement of processes and the development of tacit knowledge and skills, enriching, at the same time, the resources and capabilities endowment of the firm (Hart, 1995). In the same vein, some scholars argue that prevention-oriented firms are, from a resource-based perspective, more competitive and innovative (Shrivastava, 1995; Russo and Fouts, 1997; Sharma and Vredenburg, 1998; King and Lenox, 2002) asserting that, in fact, innovation capabilities and pollution prevention commitment are complementary assets (Hart and Dowell, 2011). Nevertheless, we think that the definition of environmental process innovations has, to date, been insufficiently addressed by environmental management scholars. As a consequence, most cited definitions are not very clear and explicit in describing the distinctive features of this kind of innovations. Fortunately, some recent contributions have made considerable efforts to define the concept (Table 2.5).

Table 2.5 Environmental process innovation definitions

Author	Definition
Ziegler and Rennings (2004), Rennings et al. (2006)	New or modified processes aimed to reduce the environmental impact
Ziegler and Nogareda (2009)	Introduction of new ecologically sustainable internal process configurations
Chen et al. (2006)	Process innovations related to pollution prevention, waste recycling, no toxicity and energy savings

Table 2.5 (Continued)

Author	Definition
González-Benito and González-Benito (2008)	Environmental Process Innovations includes: i. Responsible separation and preparation of waste ii. Emission filters and end-of-pipe controls iii. Process design aimed at energy savings and natural resources optimization iv. Production planning and control focused on waste reduction and material optimization
Rennings et al. (2006)	End-of-pipe technologies: Technologies not essential to the process and oriented to comply with the environmental legislation
Frondel et al. (2007)	End-of-pipe technologies: Technologies that reduce the polluting emissions by the implementation of additive measures
Glavic and Lukman (2007)	End-of-pipe technologies: Measures that involve the treatment of pollutants at the end of the process when all wastes have already been generated and are discharged through a flue stack or similar device
Del Río González (2005)	End-of-pipe technologies: Those devices added to production processes that, without entailing changes in the process, are aimed at transforming primary emissions into more easy-to-manage substances
Rennings et al. (2006)	Clean technologies: Those technologies that reduce the environmental impact of the operations carried out 'during the process'
Frondel et al. (2007)	Clean technologies: Technologies that reduce the use of resources at the source through cleaner production methods
Kivimaa and Kautto (2010)	Clean technologies: Technologies that are focused on process changes aimed at continuous improvement, cost minimization, resources and energy efficiency and water consumption reduction
Del Río González (2005)	Clean technologies: Process changes that aim to reduce the pollutants and wastes not only during the production process but also through the entire product life cycle (clean products)

Source: Own elaboration.

As we will see in the next section, researchers on this topic have been more interested in differentiating between so-called end-of-pipe process innovations and clean technologies.

Some contributions define environmental process innovation the same way as product innovations. It seems as if there is no difference between the two as their only distinction is that environmental process innovations are referred to as 'new or modified processes' and environmental product innovations as 'new or modified products'. Thus, according to some scholars (Ziegler and Rennings, 2004; Rennings et al., 2006), environmental process innovations are those new or modified processes that aim to reduce the environmental impact. Similarly, Ziegler and Nogareda (2009) define them as new ecologically sustainable internal process configurations.

In the same line, but more specifically, Chen et al. (2006) provides additional considerations, arguing that environmental process innovations are those related to pollution prevention, waste recycling, no toxicity and energy savings.

Therefore, as can be seen, these definitions are more results-oriented; emphasizing the final objective pursued, namely, environmental impact reduction and sustainability and energy savings or pollution prevention.

In contrast with the above, rather general, definitions, other scholars have made considerable efforts to try to specify more clearly the main features of these kinds of innovation taking a step ahead in their definition. This is the case of González-Benito and González-Benito (2008), who detail very clearly a set of process-oriented environmental practices:

i. Responsible separation and preparation of waste
ii. Emission filter and end-of-pipe controls
iii. Process design aimed to energy savings and natural resources optimization
iv. Production planning and control focused on waste reduction and material optimization

In short, taking into account previous considerations, we can define environmental process innovations as those new or modified processes oriented to the reduction of the environmental impact of internal operations consisting in the deployment and development of measures that affect the process itself and to its outcomes. Some examples of these measures are the process design changes aimed at reducing the natural resources used in the process and, within the measures oriented to control the outcomes of the process, the installation of filters and end-of-pipe controls and responsible waste separation and preparation.

2.3.2.1 End-of-pipe controls and clean technologies: Additional considerations

As previously noted, much of environmental management literature differentiates between end-of-pipe innovations and clean technologies[2] (Rennings et al., 2006). As stated by Triebswetter and Wackerbauer (2008), this distinction is relevant in terms of productivity, production cost, investing decisions, adaptation cost and compatibility with existing production methods. In this sense, it is argued that while end-of-pipe technologies reduce total productivity and increase production cost, clean technologies have the potential to boost productivity and reduce the production cost. Conversely, end-of-pipe technologies require less production investments, have lower adaptation cost and are more easily compatible with off-the-shelf production methods, in consequence diminishing the production risk.

Today it is generally accepted among environmental scholars that end-of-pipe measures are focused on the outcomes of the process rather than on the process itself. According to Rennings et al. (2006), these technologies are more oriented to comply with the environmental legislation and are not essential to the process. Frondel et al. (2007) argue that end-of-pipe technologies do reduce the polluting emissions, while according to Glavick and Lukman (2007) these measures involve the treatment of pollutants at the end of the process when all wastes have already been generated and are discharged through a flue stack or similar device.

Therefore, following Del Rio Gonzalez (2005), end-of-pipe technologies are those devices added to production processes that, without entailing changes in the process, aim to transform primary emissions into more easy-to-manage substances.

On the other hand, the discussion in regard to clean technologies lies in whether these kind of technologies are only process-oriented or are also covering products.

Supporting the first approach, we can mention Rennings et al. (2006) who consider that clean technologies are those that reduce the environmental impact of the operations carried out 'during the process'. According to Frondel et al. (2007), clean technologies reduce the use of resources at the source through cleaner production methods and for Kivimaa and Kautto (2010), these technologies are focused on process changes aimed at continuous improvement, cost minimization, resources and energy efficiency and water consumption reduction.

Nevertheless, other contributions (Del Río González, 2005) look at the issue from a more comprehensive perspective, stressing the fact that by using clean technologies these process changes aim to reduce the pollutants and wastes not only during the production process but also through the entire product life cycle (clean products).

In contrast we argue, following Glavic and Lukman (2007), that it is more appropriate to place clean technologies inside the production process so that we can differentiate between clean technologies and pollution prevention. This way, while pollution prevention covers not only the process but also environmental organizational or administrative innovations and services, clean technologies are focused on the production process.

2.3.3 Environmental organizational innovations

Since environmental management systems' main aim is the organizational change towards more sustainable management practices, they used to be considered as the typical environmental administrative innovation (Rennings et al., 2006: 48). As such, these innovations consist of the settlement or modification of firm policies, targets and responsibilities aimed to reduce the environmental impact. Through their implementation, firms get involved in a self-regulation effort oriented to adopt specific management practices that integrate the natural environment into its decision-making process and set new organizational structures to gather information and follow the environmental progress of the firm (Khanna and Anton, 2002: 541).

These environmental management standards, ISO 14001 and EMAS, are the most widespread around the world, with 200,000 firms certified in ISO 14001 in 2008 worldwide (Heras, 2011). They are oriented to reach significant improvements in the environmental performance of organizations (Khanna and Anton, 2002: 541) and also to reinforce the competitive position, image and reputation of the firm (Delmas, 2001; Wagner and Schaltegger, 2004; Darnall, 2006; Rennings et al., 2006). Through the implementation of environmental management systems, the firm can settle its environmental policy and general philosophy, signalling the environmental targets and creating the organizational structures needed to reach the goals (Darnall and Edwards, 2006). Further, with these systems it is possible to track the discrepancies between the goals and the real environmental activities, leading to appropriate corrective actions and providing firms, at the same time, with environmental legitimacy in the fight against environmental degradation.

The implementation of these systems may have interesting implications for both process and product innovations. On the one hand, environmental management systems encourage firms to take steps towards prevention, involving line employees on continuous improvement and promoting the accumulation of tacit knowledge and skills. In this sense, it can be argued that firms that have not implemented these standards may find themselves in a disadvantaged position due to the lack of guidance in the development of systematic prevention-oriented processes and skills (Hart, 1995).

On the other hand, environmental management systems also play a role in the development of product innovations. Although some scholars have mentioned that these standards are not strongly linked to product design (Rennings et al., 2006) they can help to improve the environmental awareness within the firm. Also, some additional arguments can support the importance of environmental management systems on the development of environmental product innovations. As the empirical evidence shows, most of the environmental product innovations have been carried out by certified firms (Rehfeld et al., 2007) and the incorporation of eco design aspects in the environmental management systems can lead to cost and time to market reductions (Donelly et al., 2006).

From the RBV, as argued by Delmas (2001), environmental management systems can be seen as a way to improve management quality. In this sense, they involve the entire organization, management in particular, provide coordination in the search of the environmental targets and can lead to the efficiency of operations and competitive advantage achievement. Among the strategic benefits derived from these systems we can underline processes' continuous improvement, risk reduction and competitiveness enhancing (Darnall, 2006).

There are also other environmental innovations with the same organizational character such as life cycle analysis or environmental labelling, which are characterized by having a more operative approach and are considered, in general terms, as environmental management tools that are not included in the environmental management systems we are referring to (Wagner, 2008; ISO, 2009; Ziegler and Nogareda, 2009).

As previously cited, most recognized environmental standards are ISO 14001 and EMAS. The ISO 14000 family of standards has been designed to provide an international framework for the measurement and assessment of environmental management. They do not set specific environmental performance targets but in return provide organizations with the tools to control and assess the environmental impact of their activities, products and services. Particularly, ISO 14001 is included in the ISO 14000 family and its main function has to do with the systematization of the environmental management of the company. Its implementation comprises the next five stages (Spanish Association for Standardization and Certification [AENOR], 2010):

i. Definition and environmental policy settlement.
ii. Environmental management planning: Environmental and legal requirements. Environmental targets and programmes.
iii. Implementation: Required resources, functions and responsibilities. Training, communication, documentation, operative control and emergency responses.
iv. Verification: Monitoring and measurement. Internal auditing, preventive and corrective actions.
v. Revision: Management revision.

In turn, EMAS is the European Union voluntary standard. Although in the very early stages it was designed for firms belonging to industrial sectors, in 2001 its scope was enlarged to cover all king of organizations, including those related to the government. Its main aim is to establish and implement environmental policies and programmes, to carry out continuous and systematic monitoring activities and the provision of information to the general public.

EMAS, besides including all ISO 14001 requirements, is also demanding periodical environmental information through its environmental statement, a public document that must include guaranteed and reliable information about the environmental behaviour of the firm and its environmental performance. This statement must include at least (AENOR, 2010):

i. Firm overview: Activities, products and services.
ii. Brief description of its environmental management policy.
iii. Firm's activities' direct and indirect environmental impact. Improvement objectives.
iv. Environmental behaviour in comparison with previous objectives.
v. Name and number of environmental verifier and validation date.

Therefore, according to what we have just presented, we can draw the next conclusions:

i. Environmental management systems are environmental administrative or organizational innovations.
ii. Other environmental management tools such as life cycle analysis or environmental labelling are complementary in nature but are not classified as environmental management systems.
iii. Environmental management systems involve the whole of the organization and provide potential strategic benefits

2.3.4 Environmental product innovations

Although some authors argue that the greatest environmental impacts are caused by the use and discharge of products (e.g. CO_2 emissions or metals from batteries), rather than its production, there is little literature on environmental product innovations (Kammerer, 2009). Nevertheless, some scholars have made remarkable efforts in defining these innovations (see Table 2.6), efforts that must be emphasized due to the difficulty of distinguishing, in some cases, between product and process innovations.

It is commonly accepted that environmental product innovations are those new or modified products that aim to reduce the environmental impact (Rennings and Zwick, 2002; Ziegler and Rennings,

Table 2.6 Environmental product innovation definitions

Author	Definition
Rennings and Zwick (2002), Ziegler and Rennings (2004), Rennings et al. (2006)	Innovation which consists of new and modified products with the aim of minimizing environmental impacts
Triebswetter and Wackerbauer (2008)	Innovation which comprises the introduction of new technologies, using new technologies for new applications as well as investments in new knowledge and improvements of existing products by new materials
Chen et al. (2006)	Product innovation related to energy saving, pollution prevention, waste recycling, no toxicity, and green product designs
Ziegler and Nogareda (2009)	Introduction of a product environmentally improved in the market or a new product that is environmentally sustainable
Lenox et al. (2000), Pujari et al. (2004), González-Benito and González-Benito (2008)	Innovation which lies in replacing pollutant and dangerous materials, reducing the use of resources and generation of waste during the production and their utilization and making product dismantling at the end of useful life easier, its reutilization and its recyclability
Kammerer (2009)	Innovation which reduces environmental impact during the life cycle of the product
Fiksel (2001)	Environmental product innovation characterized by the following dimensions: design for recycle, reutilization and dismantling, design for reducing wastes, design for energy saving, design for materials saving and design for the reduction of risk and prevention
Dangelico and Pujari (2010)	Environmental product innovation includes several aspects directed towards three main issues (materials, energy and pollution)

Source: Own elaboration.

2004; Rennings et al., 2006). Although very often cited, this definition is far from being specific or detailed as it does not mention how or when we can find these new or modified products inside the production process.

In the same vein, also using a generic and not very detailed perspective, Triebswetter and Wackerbauer (2008) argue that the term 'environmental product innovation' includes the introduction and use of new technologies for new applications as well as new knowledge investments and new product improvements through new materials utilization.

Chen et al. (2006) go one step further when defining this kind of innovation as those related to energy conservation, pollution prevention, waste recycling and green product design. Although a very valuable effort, it seems that all these concepts overlap one another. Thus, green product design can result in energy savings, recycling or can even be understood as part of pollution prevention strategies. And taking a completely different approach Ziegler and Nogareda (2009) constrain the scope of the term, emphasizing the role of the market. According to them, environmental product innovation happens when a new or more sustainable product is introduced in the market (e.g. cars or washing machines).

In contrast, other contributions are characterized by using a comprehensive but at the same time very specific concept of environmental product innovation. Among them, we can highlight scholars like Lenox et al. (2000) and Pujari et al. (2004, in González-Benito and González-Benito, 2008), who include in this innovation category activities such as the replacement of polluting and dangerous materials, resource consumption and waste generation reduction and product removal at the end of its life. Using a similar argument, Kammerer (2009) defines environmental product innovations as those that reduce the environmental impact throughout the product life cycle (e.g. product toxic materials reductions, product life enlargement and obsolete product recycling).

Fiksel (2001, in Tien et al., 2005), focusing particularly on product design, identifies five dimensions like design for recycling, design for waste minimization, design for energy saving, design for material saving and design for risk reduction and prevention.

Finally, with maybe the best environmental product innovation definition and classification to date, Dangelico and Pujari (2010) remark that these innovations can be reduced to three specific dimensions: materials, energy and pollution (Figure 2.2). In regard to the first dimension, the utilization of recycled or biodegradable products and packaging materials is mentioned. As part of the second

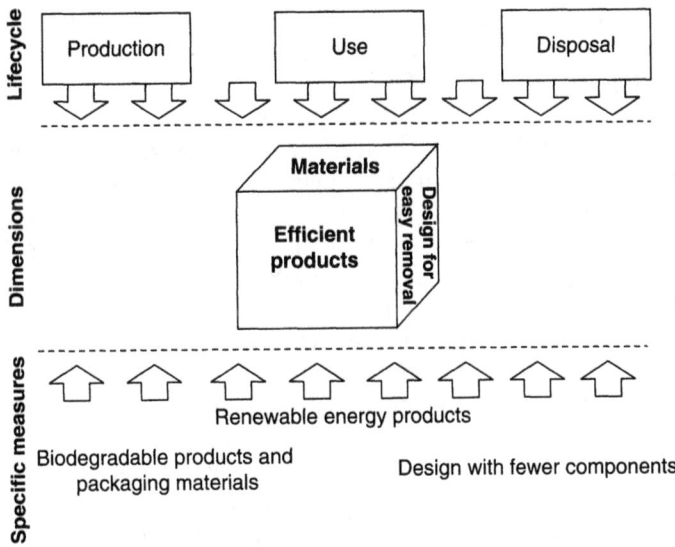

Figure 2.2 Environmental product innovation dimensions
Source: Elaborated from Dangelico and Pujari (2010: 472).

dimension, product energy consumption, we can find the design and development of efficient and renewable energy products and, finally, in relation to the third dimension, pollution prevention, activities like the product design with fewer components, aimed to achieve easier product removal at the end of its life, are also of great importance.

Together with Dangelico and Pujari (2010), we argue that product life cycle approaches are the best to explain the content and meaning of these kind of innovations. Environmental product innovations are those innovations which primarily affect the design of the products and are aimed at reducing its environmental impact throughout their production, use and disposal at the end of its life.

2.3.5 Environmental tools as supporting factors

The implementation of the above-mentioned eco-innovations can be sometimes supported by a number of environmental tools. As argued by Olcese (2008) their utilization will depend on several factors such as a firm's policy, resource and capabilities availability, normative framework to be applied or the nature of company's interactions with the natural environment. These tools can be classified in the following way:

- Organizational and management tools like environmental management systems, corporate environmental reporting and life cycle management.
- Marketing and communication tools like eco-labels, ecological marketing and environmental reporting.
- Design and product development tools like eco-design, eco-efficiency analysis, life cycle analysis, environmental risk assessment and integrated product policy.
- Supply and purchasing tools like environmental supply chain management.
- Production and distribution tools like eco-efficiency analysis, industrial ecology and life cycle cost assessment.
- Environmental tools oriented to facilities, projects and services like green building design and environmental impact assessment.

Below, we describe some of the most widely used tools: Life cycle analysis, Eco-design, Ecological label and Ecological marketing.

The life cycle analysis is the environmental tool that allows us to assess and identify the environmental impact of a product throughout its life cycle phases. Once the environmental impacts have been determined, the life cycle analysis' (LCA) main aim is to reduce the environmental impact by incorporating a number of environmentally beneficial measures. Therefore, starting from the idea that the environmental impact of a product is produced along several phases or stages like raw material extraction, production, transportation, distribution, consumption and elimination, in the LCA every environmental impact associated with a product is analysed, including the energy and materials utilization and waste and emissions generation.

Some institutions have contributed to the development of the LCA. Particularly, the ISO international regulations for life cycle analysis and the Society of Environmental Toxicology and Chemistry (SETAC).

According to Ludevid (2000, in Claver-Cortes et al., 2011), the main objectives of life cycle analysis (LCA) are as follows:

- The establishment of a firm's product and innovation policy. In this way, LCA provides information in order to decide which product has to be developed as it allows taking into consideration the financial, marketing and technological considerations.
- LCA is also a good way to detect where the environmental problems are. Once these are detected, the firm can act to overcome them.
- Using a broad and value chain oriented perspective, LCA also makes it possible to obtain suppliers' specific environmental information. It also looks at the environmental impact derived from product consumption, information that is very valuable in terms of product design and product usage recommendations. In short, LCA allows us to know in detail the environmental impact of the product at the end of the cycle. Furthermore, as the influence of LCA goes beyond the firm and involves all the agents in the value chain, it also enables business associations and different firms to work together to reduce the environmental impact, making possible as well the creation of industrial ecology systems.
- To analyse objectively the environmental improvement of the product with respect to competitors' products or substitutive ones.
- And finally, with the LCA it is easier to access the ecological label. Following Claver-Cortés et al. (2011), the LCA comprises broadly the following steps: objectives and scope determination, inventory analysis, impacts assessment, interpretation of results and improvement assessment. Below, we describe very briefly the main features of every step, noting that these steps or phases are not isolated but interrelated and that sometimes it is also possible to carry out this analysis without following every step as it will depend on the depth of the analysis that the firm wants to make.
- Objectives and scope determination: Besides the objectives, the preliminary planning and the scope of the analysis, it also includes the reasons and the expected information to be recovered. Other aspects such as the communication and the use of the information are also important.
- Inventory analysis: In this phase, a detailed inventory of all environmental loads is established for each phase of the life cycle. So, a balance of materials and energy is realized by collecting data and doing different calculations in order to quantify the entries (raw materials, energy, water and other resources) and the outflows

(waste and emissions) of the system under analysis. Other parameters such as noises, vibrations and land use among others can also be included in the analysis.

- Impact assessment: Once the inventory analysis has been executed, it is necessary to interpret it, examining and assessing the impacts produced by previously identified environmental loads. In this sense, as not all the environmental loads have the same importance and are equally harmful, it is necessary to make a classification of them according to their environmental impacts. This assessment is executed in several steps: classification, characterization and evaluation. One common classification method is to differentiate between entries (resources consumption or energy) and outflows (polluting emissions or noise, for example). Once the classification has been done, a reference point within every classification must be taken through the characterization phase. For example, inside the polluting emissions category, we must assess which polluting substance is the one that has a better explanatory power. In other words, we have to set a reference point for each category. Finally, in the evaluation phase the relative importance of previous categories and the gathered information are analysed according to several criteria taking into account not only the economic but also the socio-economic aspects.
- Interpretation and improvement assessment: Finally, impact assessment must be carried out. In this last step, the main aim is to provide conclusions and recommendations in line with the objectives and scope of the study, taking into account the results of previous phases. These conclusions as well as the identified environmental improvement options of the product will be particularly important to the eco-design elaboration.

One particularly interesting example of the application of the LCA is the case of the 3M Corporation (Olcese, 2008). With the main aim of improving the environmental performance of the company, 3M has implemented the LCA in most of its new product development process worldwide. Three main factors were critical for LCA implementation. First, there was a strong support of the organization and several teams were involved in the analysis (Corporate Product Responsibility team, 3M Technological Centres, Environmental, Safety and Health Corporate Committee). Second, four different impacts were analysed (environment, energy resources, health and security) and three different phases in the product life cycle (materials supply, R&D operations, production). Third, not only is the economic value taken into consideration but also other value categories like intangible values or long-term values among others.

We understand eco-design as the way in which design function can improve the environmental impact of the firm. Usually, eco-design is referred to product design, although some aspects of productive process design can be also included within the term eco-design. Thus, using a LCA perspective, the ultimate goal of eco-design is also, through the design of products and processes, to reduce environmental impact along the phases of the value chain, from raw materials extraction until the end of useful life, including manufacture.

So, environmental design of products consists on the incorporation of environmental criteria in product conception and development, attempting to anticipate future environmental impacts. Those impacts are referred to production phase and consumption and waste conversion phases, as well as impacts produced in previous stages of the life cycle, such as energy production and distribution, raw materials extraction and elaboration, and manufacture of components.

In terms of environmental strategy, eco-design supposes a step towards environmental sustainability. Leaving behind the pollution control and pollution prevention strategies, eco-design means, as Hart stated in 1995, incorporating the 'voice of the environment' in the design of the product using a product stewardship strategy, that uses a typical life cycle analysis philosophy, involving all the value chain participants.

Furthermore, in the same NRBV, eco-design can also improve the financial performance of the company. Among others, the beneficial effects of the eco-design are (Ludevid, 2000):

- To add value to firm's products portfolio. Incorporating the environmental criteria into product design and development is an indicator of the innovation capability of the firm.
- To contribute to product differentiation with respect to competitors, paying attention to those market segments particularly concerned with environmental problems.

- Eco-design is also a sign of environmental quality that provides image and marketing advantages that in the end can be decisive to obtain the ecological label.
- Eco-design can be a useful method of environmental communication.
- Through the eco-design it is possible to reach important cost reductions.
- Eco-design is also a useful way to avoid customers' complaints.
- Eco-design is also a way of getting ahead of environmental regulations. That is, firms implementing eco-design principles will have lower adaptation costs to environmental regulations.
- Eco-design can also be a business opportunity if the firm exploits and sells its acquired knowledge.

After presenting a brief overview about life cycle analysis and environmental or eco-design of products and processes, now we focus on ecological label and ecological marketing.

Ecological label or eco-label is a voluntary sign awarded to those products that have low environmental impacts within the different phases of the life cycle (Ludevid, 2000). In order to get a certified eco-label and be able to use the logotype in the market, an independent evaluator has to check that the product fulfils the ecological criteria previously defined, depending on the type of product. This requirement, the external verification, gives the eco-label a sign of authenticity and makes possible the product differentiation.

The benefits of the eco-label can be separated into two different groups: On the one hand, the eco-label is a good form of promotion as the firm can show outside that its products fulfils the environmental, quality and security requirements. Furthermore, the eco-label not only benefits the product but also the company as it shows an image of corporate responsibility. On the other hand, eco-label also benefits the customers as it provides more and better information to them. In this sense, the customers can count on the environmental quality of the product they are using.

There are many countries with different eco-label systems (Seoánez and Angulo, 1999), such as Blue Angel (Germany), European eco-label (European Union) or Environmental AENOR (Spain). Their methodologies are quite similar and not all the products can be awarded with the eco-label. If a firm wants to attain an ecological label for its product, it should first check that the product belongs to the categories that have the possibility to obtain the eco-label. The different products are classified into groups or categories and each one of them has specific ecological criteria to be fulfilled. Once the company finds the category where its product belongs and the environmental criteria are fulfilled, the firm should submit an application at an authorized organization. As we have noted before, the eco-label needs external verification.

Finally, we will analyse briefly some aspects of the ecological marketing as an additional tool that supports the environmental efforts of the company. The ecological marketing concept finds its reason in the emergence of the so-called 'green consumers' or 'green consumerism' movement. The main feature of this group of customers is that they are deeply worried about the impact of firms' manufacturing activities on the environment. They want to know how strong is the environmental impact of the products they buy, not only in manufacturing terms but also in regard to other aspects like energy used and waste generation among others. Furthermore, these customers could also be willing to pay an additional price in consideration of the environmental qualities of the product.

Although is true that the environmental concern among the customers is far from being generalized, it is also true that these kind of customers can be a niche market for the companies. In this sense, firms can carry out strong environmental marketing strategies directed towards this kind of customer. That is the reason why the environmental or ecological marketing has emerged.

Thus, ecological marketing can be understood as the utilization of the marketing variables (product, price, distribution and communication) in order to satisfy customers' needs, achieve firms' objectives, and generate the minimum negative impact on environment. Some of the environmental marketing activities in relation to the marketing variables can be as follows (Claver-Cortés et al., 2011):

- Product level: Product design that minimizes the environmental impact thorough all life cycles, having in mind that the desired ecological function should not affect negatively other attributes of product differentiation. At this level, the previously analysed eco-label can play an important role.
- Price policy level: The objective is to establish a price which captures a firm's cost structure. Therefore, it is necessary to know how much a customer would be willing to pay for an ecological

product. It is assumed that environmentally deeply concerned customers would be willing to pay an additional price for the ecological dimension of the product.
- Distribution policy level: This aims to minimize the resources and the generation of waste during physical distribution (transportation, storage and operation), as well as incorporation of the environmental factor within the decision-making process of the wholesaler.
- Communication policy level: Ecological communication should be focused on environmental education, contributing to create an image of environmental responsibility.

2.3.6 Other approaches to environmental innovation

Besides the environmental or eco-innovation classification proposed by Rennings et al. (2006), which constitutes our main reference, there are other contributions that analyse the eco-innovation phenomenon from sometimes very different and sometimes complementary perspectives. As argued by Carrillo-Hermosilla et al. (2010), one could create a classification attending to the environmental performance or to the environmental motivation of these innovations. Nevertheless, this would be a difficult task, as sometimes the environmental performance determination is far from being clear and it is not easy to calculate. The same can be said in regard to the environmental motivation of eco-innovations. Thus, some innovations that are not catalogued as 'environmental' can, in the end, generate beneficial effects for the natural environment and conversely, the effects of other environmentally friendly innovations will depend on many factors (how these innovations are deployed, how these innovations interact with other incumbent measures, etc.).

Therefore, the below additional approaches to eco-innovation can either be catalogued as environmentally motivated or not, can improve the environmental performance of the firms or, on the contrary, can even cause more damage to the environment. For example, as we will see in the next section, some general purpose innovations in the end become essential for environmental improvement.

Although the most commonly used eco-innovation classifications deal with the technological versus organizational issue (Kemp and Foxon, 2007), there are also other alternative classifications that pose interesting perspectives. In this sense, while some are more focused on evolutionary approaches (Carrillo-Hermosilla et al., 2010), others (Andersen, 2008) propose a classification based on eco-innovations' level of impact on both the environment and the economy in general. In the next section we will briefly review some of these approaches.

The four dimensions of eco-innovations. According to Carrillo-Hermosilla et al. (2010), eco-innovations can be also classified in a different way. These authors contribute to the literature on environmental innovation offering a distinctive classification based in four dimensions called design, user, product service and governance. Based on an evolutionary perspective of innovation and highlighting the fact that 'innovation arises through a systemic process that refers to the interconnectedness and dynamic interaction between different actors and internal and external factors influencing the innovation process' (Carrillo-Hermosilla et al., 2010: 1075), they argue that their four-dimension classification covers most environmental innovations and contributes to a better understanding.

The first dimension they mention is the Design Dimension, which is further divided into three different classes called Component Addition, Sub-system Change and System Change.

Component addition refers to those devices added to the process that improve the environmental quality without changing the nature of process that is generating the environmental impact. These devices or components are what we have previously called end-of-pipe measures. As these authors mention, the implementation of these measures do not impact the main process and are used by firms with two purposes: (i) as a way to 'gain time' before introducing new and more efficient technologies and (ii) to comply with environmental regulations.

Sub-system change or eco-efficiency measures aim to do more with less or create more products and services with fewer resources. Although they represent a step in the right direction, these kind of measures are very limited as they are restricted only to the firm and its impact is quite limited.

System change measures are based on a more comprehensive industrial ecology logic that tries to create industrial processes that mimic nature's own processes. They represent the more environmental committed closed loop systems approach where the wastes of previous processes become the inputs of new processes.

The second dimension is called the User Dimension. It emphasizes the role of customers in eco-innovations from two different perspectives. On the one hand, it is highlighted that firms, sometimes, assuming that some customers are very creative, try to involve them in order to benefit from their ideas in the product and service development phase. On the other hand, customers are always a crucial factor as the acceptance of the product or the service at the end depends on them. This argument can also be valid for environmentally friendly products but with one additional handicap as the market for this kind of products is far from being consolidated.

Product service dimension focuses on the eco-innovation value chain. It is not only the firm, as eco-efficiency proposes, but also the stakeholders that play a role in the product or service perception, development and delivery. This dimension is, in a sense, very close to the concept of product stewardship developed by Hart in 1995 as it involves the entire value chain (supply, delivery and customers) in the search for more sustainable solutions. The creation of these sustainable value chains will in the end lead to the generation of competitive advantages to the firm, creating a network that enables it stay ahead of competitors.

Finally, the governance dimension refers to those changes (mainly at an institutional or system level) that are essential to eco-innovation success. Sometimes, the technology is already developed but there are many barriers (both at firm and government level) that prevent its implementation. When this happens, only certain governance initiatives at institutional level can unlock the innovation process. At firm level sometimes it can be necessary to create new partnerships or relationships with other firms or agents and at government level it may be crucial for the settlement of the right conditions to promote or stimulate the development of these eco-innovations. This latter point is particularly important as the economic benefits from eco-innovations are hard to get and the incentives to invest in them are sometimes reduced. As mentioned by Rennings (2000), the double externality problem means that besides the positive externalities common to all innovations, eco-innovations also generate positive externalities via the external environmental cost reductions. This means that although the cost of innovation will be supported by a single firm, the benefits will be enjoyed by the entire society. Furthermore, the firm will find that its ecological products are at a clear disadvantage in terms of price compared to other non-ecological alternatives.

Preventive technologies. Other well-known environmental innovation taxonomy is provided by Hohmeyer and Koschel (1995, in Rennings, 2000). This approach is in a sense similar to the previously mentioned system changes in the Design Dimension as it reflects a closed loop perspective where the outputs of the processes and the outcomes from production and consumption are the inputs for subsequent stages.

According to their point of view, environmental technologies can be separated into two main levels, integrated technologies and additive technologies. The integrated technologies are focused on the production process involving inputs, production process and outputs (products) and additive technologies, which are more focused on the effects and emissions of production and consumption.

Integrated technologies and additive measures. Among the measures focused on the inputs of the production process we can find, for example, the substitution of ecologically harmful inputs. Production process measures refer to the optimization of process components, the integration of new or alternative components in the process or the use of alternative production processes. Finally, in regard to the measures focused on the outputs of the production process, the main focus is the product, more specifically, the optimization of single product components, the integration of new product components, and the exchange of single product components or complete products. It is worth noting that although Rennings (2000) seems to equate additive protection and end-of-pipe measures, it could be argued that the devices installed inside the production process aimed to recover some emissions and send it back to the process could be catalogued as end-of-pipe as well.

Finally, according to this classification, additive measures are understood as external processes. Among them we can find those curative processes like soil decontamination and secondary (out of the process) recycling.

Environmental technologies, organizational innovation and new technology systems. Kemp and Foxon (2007), in turn, offer three categories for environmental innovations called environmental technologies, organizational innovations and new technology systems.

One interesting point in this classification is that the third category is a combination of the previous two.

In order to be more specific, the author carries out a remarkable effort detailing the eco-innovations included in each category. Thus, among environmental technologies we can find pollution control technologies (including waste water treatment technologies and cleaning technologies that treat pollution released into the environment), cleaner process technologies or new production processes, waste management equipment, environmental monitoring instrumentation, green energy technologies, water supply and noise and vibration control.

Organizational innovations, in turn, are understood in a comprehensive way as they include not only intra- but also supra-organizational changes. Therefore, it can be appreciated that this classification is more or less in the same line as the governance dimension of eco-innovation previously noted. Among organizational eco-innovations, Kemp includes the pollution prevention schemes that are focused on input substitution, processes and production plants (note that products are not included), environmental management and audit systems (defined as 'formal' systems) and chain management understood like cooperation between firms in order to avoid environmental damage in the value chain.

The third category encompasses the previous two and is called new or environmentally improved products, environmental services and new technology systems. It refers to environmentally improved products and services derived from the implementation of environmental technologies and organizational innovations. Among them we can find environmental consulting or analytical services. This category also includes the sub-category of new technology systems which puts together the technology, the knowledge, the institutions, the infrastructures and even the changes in consumer behaviour that lead to environmental innovation success.

It can be appreciated as a distinctive feature of this classification the treatment given to the products as it seems to be outside the environmental technologies and the pollution prevention schemes mentioned as organizational innovations.

Add-on, integrated, alternative, macro and general purpose eco-innovations. Andersen's taxonomy (2008) also contributes to explaining and understanding better the great diversity of eco-innovations, emphasizing the importance of the interaction effects among them or the eco-innovation dynamics behind them.

The first group of this taxonomy includes the add-on eco-innovations. It refers to the artefacts or services that improve the environmental performance of the customer. With this generic expression, the author refers to the wide variety of technologies aimed to clean up, recycle, measure, control, transport, extract or supply natural resources and energy. The effect of these innovations in terms of sustainability is limited as they keep almost intact current production practices (business as usual) and life standards. Filters and pollution control devices are included in this classification.

Integrated eco-innovations constitute the second group. This kind of innovation aims to gain eco-efficiency, improving the overall environmental performance of the company or the environmental impact of its products. It includes cleaner technological processes and cleaner products that contribute to change production and consumption practices within the organization. Fundamentally technological, they can also involve organizational changes. Although they represent a step towards sustainability and are more environmentally committed than the add-on eco-innovations, their impact is also quite limited as they are based on incremental changes rather than radical ones. New process design or step reduction within the process can be catalogued as integrated eco-innovations.

The third group of eco-innovations is called alternative product eco-innovations. Instead of the continuity of integrated eco-innovations, they represent a radical change. As they can be considered radical product innovations they offer more environmentally friendly solutions and also a very different approach that claims for real changes both inside the firm and also in the consumption patterns of the customers. The renewable energy technologies or the organic farm falls into this category. One good example of the importance of alternative product eco-innovation is provided by Schilling (2008) when referring to hydrogen fuel-based technologies. As hydrogen is the most abundant resource available on earth, many ecologists and environmentalists have pinned their hopes on the development of this kind of technology. Hydrogen/fuel cell vehicles production represents a radical change both for firms, customers and even governments. For firms, it will be necessary to carry out a radical shift towards new production methods and to the promotion of new basic skills and capabilities among the employees. On the customer side, it is clear that the commercial success of the future hydrogen vehicles will depend to a great extent on the ecological commitment of the customer as these models will probably be more expensive than current alternatives. Finally, this particular case will be also a challenge for governments as it will be necessary to create a complete fuel supply infrastructure.

Macro-organizational eco-innovations represent the fourth category within this classification. Their impact is even greater than previous one as they pursue new and more environmentally benign ways of organizing society. These cannot be catalogued as technical or organizational and their impact goes beyond the firms, involving particularly governments and public authorities in general. Their radicalism is not on the technical side but in the concept itself.

The last category within this classification is represented by the general purpose eco-innovations. This category is also one of the most remarkable contributions to the eco-innovation literature as it is almost absent in most eco-innovation taxonomies. The main feature of these innovations is that they do not have to be environmentally oriented. Although conceived with another purpose, their impact is so deep that it affects many other fields due to their versatility and importance. Biotechnology or nanotechnology advances are among the most cited innovations of this kind.

2.4 Green corporate reputation and image

In accordance with the general purpose of this book, a clear delimitation of corporate reputation is needed. Following the same line as Fombrum and Shanley (1990), Frooman (1999) and Olmedo Cifuentes (2011), we think that social sensibility is closely linked to corporate reputation, being at the same time one of its main dimensions. Therefore, the concept of corporate reputation is the touchstone of our analysis, as in the remainder of the book we will focus on one of its dimensions, corporate image, whose connection with reputation will depend on the industry under investigation (Frooman, 1999).

Generally speaking, a good corporate reputation is recognized as one of the main organizational factors responsible for a firm's sustained competitive advantages and financial performance (Roberts and Dowling, 2002). This strategic potential is due to its own value-creation capability and to its intangible character, that is quite hard for competitors to imitate, allowing maintaining a competitive superior position, as well as differentiating the firm from its competitors and reinforcing the firm's efforts in social and environmental sustainability.

In this section we will review the concept of corporate reputation from a strategic management perspective in order to highlight its main distinctive features and shed some light on its delimitation from other similar conceptual approaches like green image and corporate identity. Then we delimitate corporate reputation's main dimensions, particularly focusing on the social and environmental one.

Finally, we will describe the concept and characteristics of green and social corporate reputation as well as its close connection to firm technological environmental product innovation and firm performance. These are connections that, from our perspective, can be extremely important for the development of previously mentioned product stewardship strategies.

2.4.1 Corporate reputation. Review, proposed definition and characterization from a management perspective

Nowadays, we can state that business management researchers and especially researchers from the RBV (Barney, 1986, 1991; Dierickx and Cool, 1989; Grant, 1991; Deephouse, 2000) have highlighted its strategic nature, helping firms to obtain sustained competitive advantages, through the reinforcement of product differentiation competitive strategies, leveraging other organizational resources and capabilities or reducing agency cost, among others. Nevertheless, corporate reputation as a business complex phenomenon needs clarification of its conceptual delimitation and characterization (Martín-de Castro, 2008).

In order to conceptualize corporate image, reputation and other similar intangible realities (whose distinction will be treated later on), a useful landmark is the work of Shenkar and Yuchtman-Yaar (1997), which provides a preliminary synthesis and comparison of how the term has been treated from such different fields as sociology, marketing, law, accountancy, economics and business management. Table 2.7 provides a summary.

Each of these disciplines implies different approaches, types of organizations studied and expected consequences of the reputation phenomena. From sociology, the focus is on the organization's status and social aspects of a variety of organizations – military, educational or non-profit organizations. Nevertheless, the remaining approaches – marketing, accountability, economy and management – are focused on a specific type of organization: the firm. Another key issue is related to the labelling one. While sociology employs the generic term of prestige, useful for all types of organizations and institutions, as well as to individuals, marketing uses image and brand terms, more linked to firm advertising expenditures, and with a specific type of key informants – customers. Furthermore, both accountability and economy use a relative defined term labelled 'goodwill', focused on its economic assessment. Finally, many management scholars use corporate reputation that is built on the basis of the assessments of multiple key informants.

The diversity of approaches used to address these similar terms requires a detailed analysis in order to classify the existing multiple definitions (Fryxell and Wang, 1994; Shenkar and Yuchtman-Yaar, 1997; Deephouse, 2000). As previously noted, among management scholars, the term 'corporate reputation' has been widely used. According to its importance, in the next section, and taking as main reference point the work of Barnett et al. (2006), we will make an actualized typology of corporate reputation conceptualizations depending on whether we understand corporate reputation as awareness, as an assessment or as an asset.

In that sense, a set of proposals understand corporate reputation as 'awareness' or 'a collective awareness' about the firm made by its different stakeholders, both external and internal (see Table 2.8).

Another set of authors understand corporate reputation as an assessment made by an extensive group of external and internal firm stakeholders. This point of view implies a certain firm's judgement in different ways made by them (see Table 2.9).

Finally, within a third stream, we can find proposals for corporate reputation as an organizational factor, asset, resource or capability,

Table 2.7 The treatment of corporate image and reputation from different disciplines

	Sociology	Marketing	Accountability and Law	Economy	Management
Term	Prestige	Image	Goodwill	Goodwill, reputation, image	Corporate reputation
Unit of analysis	Job, industry, organization	Brand, organization	Organization, individual	Organization, Individual	Firm
Studied organizations	Education, military, non-profit organization	Firm	Firm	Firm, product	Firm, group of firms
Key informants	Expert panel, partners	Potential and actual customers	Acquiring firm	Advertising expenditures, product signalling	Managers, industry and financial annalist
Reputation effects	Status, organizational ties, recruitment	Product acquisition by customers	Firm's earnings, price of merger/acquisitions	Product price, differentiation and quality	Investor's attractiveness, customer's trust on firm products, competitive advantage

Source: Adapted from Shenkar and Yuchtman-Yaar (1997).

Table 2.8 Reputation as 'collective awareness'

Author	Concept
Olmedo Cifuentes (2011)	Firm's whole and stable assessment that is shared by multiple stakeholders
Martín de Castro et al. (2009)	The result of a firm's legitimating process made by different stakeholders evaluating multiple firm's aspects and actions
Brown et al. (2006)	Organization's perceptions made by its external stakeholders
Rindova et al. (2005)	Stakeholder's perceptions on firms' ability for value creation
Larkin (2003)	Reflection of a (firm's) name
Pharoah (2003)	Exist in the eye of the beholder
Einwiller and Will (2002)	Net perception
Roberts and Dowling (2002)	A perceptual representation of a company's past actions and future prospects
Balmer (2001)	Latent perception of the organization
Fombrun (2001)	Collective representation of past actions and future prospects
Fombrun and Rindova (2001)	A collective representation of past actions and results
Hanson and Stuart (2001)	The corporate image over time
Zyglidopoulos (2001)	Set of knowledge and emotions
Ferguson et al. (2000)	What stakeholders feel and think about a firm
Fombrun and Rindova (2000)	Aggregate perceptions
Mouritsen (2000)	An ambiguous assemblage of hunches
Balmer (1998)	Firm's perception
Fombrun (1998)	Collective representations of past actions
Fombrun and van Riel (1997)	Collective representations
Fombrun (1996)	Representation based on perceptions of past actions
Weigelt and Camerer (1988)	A set of economic an non-economic attributes

Source: Own elaboration based on Barnett et al. (2006).

Table 2.9 Corporate reputation as assessment

Author	Concept
Walker (2010)	Aggregate and relatively stable representation of past actions and future prospects
Carter (2006)	Set of key firm's characteristics made by different stakeholders
Rhee and Haunschild (2006)	Consumer's subjective assessment on producer's perceived quality
Barnett et al. (2006)	Observer's collective judgements of a corporation based on assessments of the financial, social and environmental impacts attributed to the corporation over time
MacMillan et al. (2005)	Set of emotions and experiences made by stakeholders on the base of questions like: What for? Whom for? Which aims? Who for?
Dolphin (2004)	Judgement made by different stakeholders about an organization's past actions
Larkin (2003)	Value judgement
Groenland (2002)	Emotional concept based on quality, price, service, recognition and so on
Mahon (2002)	An estimation of a person or a thing
Wartick (2002)	The aggregation of a single stakeholder's evaluation
Fombrun (2001)	Collective and subjective assessment of a firm's efficiency
Fombrun and Rindova (2001)	Gauge of a firm's relative standing
Schultz et al. (2001)	Judgement of firm's perceptions and assessments
Cable and Graham (2000)	An effective assessment
Deephouse (2000)	Firm's assessment
Dukerich and Carter (2000)	Perception-based assessments
Fombrun and Rindova (2000)	General esteem
Baden-Fuller et al. (2000)	Assessment of the firm's resources and capabilities

Table 2.9 (Continued)

Author	Concept
Fombrun (1998)	Describes the whole attractiveness of a firm
Rindova and Fombrun (1998)	An aggregation of assessments made by stakeholders
Fombrun and van Riel (1997)	Subjective collective assessment/relative position of a firm
Dollinger et al. (1997)	Quality's expectation
Fombrun (1996)	Overall estimation of a firm
Brown and Perry (1994)	Firm's evaluation
Dowling (1994)	Firm's assessment in relation to (respect, esteem, estimation)
Hall (1992)	Knowledge and feelings of individuals
Fombrun and Shanley (1990)	Public's cumulative judgements

Source: Adapted from Barnett et al. (2006).

which a firm can control and manage and it adds value to the firm, that is, corporate reputation as an economic firm asset. See Table 2.10.

As can be seen in previous tables, the majority of proposals made were focused on the assessment of certain organizational attributes, and measured as perceptions by a set of key informants. In this way, one of the early relevant academic literature corporate reputation concepts was made by Weigelt and Camerer (1988), considering it a set of attributes of a firm that is the result of past actions and they can generate firm profitability.

Other authors focus their interest on a stakeholder perspective in order to give a definition of corporate reputation. Such is the case in Petric et al. (1999), who define it as the result of a competitive legitimating process where firms have to transfer their key characteristics to stakeholders in order to maximize their moral and socio-economic status. In the same line of reasoning, Deephouse (2000) understands that corporate reputation arises from the organizational assessment made by its stakeholders in terms of its influence, knowledge and esteem. That is, corporate reputation emerges by the

Table 2.10 Reputation as an asset/resource

Author	Concept
Olmedo Cifuentes (2011)	Intangible resource
Martín de Castro et al. (2009)	Organizational capability
Goldberg et al. (2003)	Intangible resource
Mahon (2002)	Firm resource/firm asset
Miles and Covin (2002)	Valuable, but fragile intangible asset
Fombrun (2001)	Economic asset
Barney (2001)	Organizational capability
Drobis (2000)	Intangible asset
Miles and Covin (2000)	Intangible asset
Petric et al. (1999)	Result of firm's key characteristics transferred to its stakeholders
Shenkar and Yuchtman-Yaar (1997)	Firm's ranking based on relevant criteria
Riahi-Belkaoui and Pavlik (1992)	Key asset
Hall (1992)	Intangible asset
Spence (1974)	The result of a competitive process

Source: Adapted from Barnett et al. (2006).

interactions between the firm and its stakeholders and by firm information and actions among them, including specialized informative intermediaries.

Finally, we would like to highlight the definition made by Roberts and Dowling (2002), who consider corporate reputation as the perception of a firm's past actions and future perspectives that describes the whole firm's attractiveness when it is compared with other rivals. This proposal suggests that corporate reputation is an organizational attribute of general character that reflects the extent to which firm's stakeholders qualify the firm as good or bad.

From a strategic management perspective, and more concretely, from a resource-based view (Barney, 1991) and considering corporate reputation as an organizational factor, Baden-Fuller et al. (2000) refer to it as the assessment of a firm's bundle of resources and capabilities made by a clearly defined set of collectives, stakeholders or the public. Thus, when we try to define and manage the main characteristics of that concept, we have to keep in mind the following issues:

- It is built during a process that includes past results as well as future ones with the ability of generating rents.

- Its assessment is made by a set of firm's heterogeneous and different stakeholders.
- By means of corporate reputation, firms achieve recognition and social legitimating.

In order to propose a definitional landscape on corporate reputation, from the review of the different contributions and framework previously described (Weigelt and Camerer, 1988; Fombrun and Shanley, 1990; Rao, 1994; Fombrun, 1996; Dollinger et al., 1997; Shenkar and Yuchtman-Yaar, 1997; Petric et al., 1999; Baden-Fuller et al., 2000; Deephouse, 2000; De Quevedo, 2001; Groenland, 2002; Roberts and Dowling, 2002), as well as from the characteristics provided by Fombrun and Van Riel (1997), we could say that corporate reputation is the result of the process of 'social legitimization' of the firm, and to define it in the following terms:

> Corporate reputation constitutes the result of a legitimating process in which certain external and internal firm's stakeholders assess multiple firm's aspects, including past actions, the bundle of organizational resources and capabilities, and future firm's performance expectations, which are necessary to firm's value creation and profitability.

Once we have reviewed and proposed an integrative definition of corporate reputation, we have to address two additional issues:

- To differentiate corporate reputation from two so closely related business concepts: corporate identity and corporate image.
- To characterize corporate reputation in order to understand its true complex nature.

2.4.2 Corporate reputation, corporate image and corporate identity: Exploring the 'terminological jungle'

Strategic management literature, due to its accelerated pace of growth, is characterized by a lot of new and evolving concepts, constructs and variables, which many authors refer to as a 'terminology

The Nature of Environmental Innovation and Green Image 73

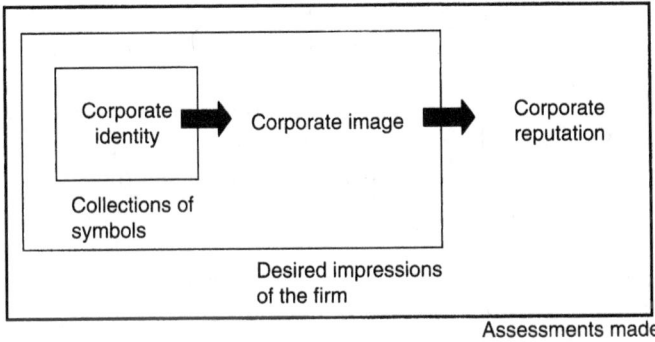

Figure 2.3 Corporate reputation, corporate image, corporate identity
Source: Own elaboration based on Barnett et al. (2006).

jungle'. That is the case in corporate reputation and its closely related concepts 'corporate image' and 'corporate identity' (Villafañe, 2004). In order to shed some light on this topic, the contributions of Minguez (2000), Barnett et al. (2006) and Olmedo-Cifuentes (2011) would be useful. Figure 2.3 offers a relational model of these concepts.

Corporate identity embodies the underlying 'core' or basic characteristics of the firm (Barnett et al., 2006). Fombrun and van Riel (2004) remark that corporate identity includes: (i) features that employees consider key for the firm; (ii) features that make the firm distinctive from others; and (iii) features that are enduring, allowing the connecting of the present and the future. In this way, corporate identity is a collection of different symbols (Olmedo-Cifuentes, 2011): visual, cultural, environmental and behavioural ones.

Informally, corporate image is what comes to mind when you hear the name or see the logo of a company. That is, it refers to the general impression that a firm generates in the mind of its stakeholders (Dowling, 2004). In that sense, corporate image constitutes the desired current general impressions of the firm to its key stakeholders, with a more short-term focus, being the firm's internal projection in order to influence stakeholders' perceptions of it. In this way, corporate image is usually constructed, modified and/or improved with several management and communication activities like publicity, advertising campaigns and so on.

Finally, corporate reputation is the result of the firm's collective assessments and judgements made by different – internal and external – stakeholders, in a legitimating process over time, adding firm's identity values, corporate image, perceptions, beliefs, experiences and so on. This aggregate and complex conception gives to corporate reputation a long-term focus, being more stable and solid over time.

2.4.3 Corporate reputation: Main features

In order to fully explore the nature of a firm's corporate reputation, and in order to do a better strategic analysis of it, we propose a set of key characteristics (Figure 2.4).

In that sense, the work of Barney (1999) highlighted corporate reputation as an organizational capability with the following main characteristics: (i) unique historical context, (ii) path dependence, (iii) social complexity, (iv) tacitness and (v) difficult to create or accumulate. The firm's corporate reputation is built or accumulated in a historical context and in unique circumstances that surely cannot be repeated. In the following section we briefly expose these characteristics:

Its construction is subject to unique historical business and environment contexts
Its construction is subject to path dependence that is impossible to replicate
It is difficult to create and accumulate, depending on multiple external and internal factors and actions
Due to its perceptual nature and the large number of assessments, it is a social complexity business phenomenon
Based on previous characteristics, it has a high degree of tacitness, mainly due to its intangible nature

Figure 2.4 Corporate reputation: Main strategic characteristics
Source: Own elaboration.

Unique historical context, path dependence and long time period of construction are characteristics so closely related that they can be jointly analysed. In this way, several years ago, Hall (1992) argued in his empirical survey among British managers that firm's reputation and product's reputation needed several years to be shaped. In the same line, the work of Martín-de Castro et al. (2009) found that good corporate reputations need more than a decade to be correctly built. These assertions make corporate reputation, jointly with corporate culture, one of the most difficult to accumulate organizational factors. Furthermore, Dierickx and Cool (1989) argued that, in order to obtain a good reputation, several management decisions about operations, quality and so on are needed, that is, this slow and long period of time construction is subject to path dependence and unique historical business and environment conditions, impossible to replicate or modify in the short term, moved by private manager interests (Deephouse, 2000).

Taking into account complex and social nature, as well as its tacitness, we can easily deduce the social and complex nature of corporate reputation, mainly due to the high number and heterogeneity of internal and external stakeholders involved in its assessment or judge – workers, managers, shareholders, customers, suppliers, allies, competitors, media, public administrations, society in general, as well as by the fact that corporate reputation is the result of these 'collective perceptions', with many possible dimensions. In this way, we can say that corporate reputation shows a high degree of tacitness and intangibility, and this makes its identification, measurement and management really hard.

2.4.4 Social and green corporate reputation

The detailed review of corporate reputation offers us a multitude of business aspects, perceptions and informants involved in its complex and social construction, which indicates its multidimensionality.

In this sense, proposals made by Weigelt and Camerer (1988), Dollinger, Golden and Saxton (1997), Deephouse (2000), Mínguez (2000), De Quevedo (2001) and Martín-de Castro et al. (2006) show in their investigations the multidimensionality of corporate reputation, as Table 2.11 shows.

As shown, one of the most discordant points refers to corporate reputation multidimensionality (Fombrun and Shanley, 1990). Maybe, as Roberts and Dowling (2002) state, this general

Table 2.11 Dimensions of corporate reputation

Authors	Dimensions		
Weigelt and Camerer (1988)	Firm reputation	Product/service reputation	Organizational culture reputation
Dollinger et al. (1997)	Management reputation	Product/service quality reputation	Financial reputation
Mínguez (2000)	Social reputation	Commercial reputation	Finance/economic reputation
De Quevedo (2001)	Internal reputation	External reputation	
Martín-de Castro et al. (2006)	Business reputation	Social reputation	

Source: Own Elaboration

disagreement could be due to the existence of 'financial halo' phenomenon. This happens because the majority of corporate reputation's respondents in rankings and media are financial analysts and managers, focused on firms' financial and economic performance, their corporate reputation assessments and judges are biased by firms' financial health. Dollinger et al. (1997) point out the multidimensionality of this construct, identifying three main components for reputation: managerial reputation, financial reputation and product reputation. These dimensions are independent and represent different aspects of corporate reputation and their role with different stakeholders, as the predominant framework in the study of corporate reputation.

Inside this framework, another interesting proposal is due to De Quevedo (2001). In her research, she highlights the existence of two main dimensions of corporate reputation: internal reputation and external reputation. The first is related to the 'business stakeholders' perception of firm activities – workers, managers, shareholders, customers, allies, suppliers – and the second is related to the external stakeholders' perception of firm activities – society in general. These findings are quite similar to Deephouse's (2000) work.

The work of Martín-de Castro et al. (2006) highlights the existence of two differentiated dimensions: business and social. The first one is developed, adding the assessments of stakeholders related with day-by-day business activities. Social reputation implies the assessments made by the society in general and reflect the implication of the firm in social issues, reflecting the process of business and social legitimating of the firm.

In a similar way, Barnett et al. (2006) explicitly include in their definition of corporate reputation the assessments or judgements made by stakeholders in terms of firm's financial, social and environmental impacts.

Previous proposals highlight the importance of environmental/social reputation as a key piece and dimension of corporate reputation. In this line of reasoning, other authors like Dolphin (2004) and Walker (2010) remark that firms have different reputations based on the assessments made by different stakeholders on business, social and environmental issues and firm activities. A similar perspective adopts the work of Brammer and Pavelin (2006) in which corporate reputation is determined by economic performance, market risk, institutional property, the nature of firm activities, as well as by firms' social performance.

In this way, firms' social and environmental reputation, as a key dimension of corporate reputation (Zyglidopoulos, 2001) has been partially studied in the specialized literature. With increasing importance, works by Brammer and Pavelin (2006) and Martín-de Castro et al. (2006) tries to link corporate reputation with corporate social and environmental responsibility.

There is an increased general agreement among academics, managers and society in general about the importance of a good responsible corporate image and reputation in social and environmental issues (Mahon, 2002; Brammer and Pavelin, 2004; Martín-de Castro, 2008), reflecting the strategic role and necessity of environmental sustainable business activities in convergent to Montreal and Kyoto International Protocol.

The seminal work of Fombrun and Shanley (1990) empirically demonstrates that firm's social sensibility – measured by the level of charity donations made by the firm – is positively linked with a firm's corporate reputation. In this sense, William and Barrett (2000) give additional empirical evidence linking philanthropy and reputation. As an interesting research output of this work, they found strong evidence of philanthropy–reputation link for firms which frequently make violations in health, security and environmental issues. That is, charitable donations have important 'healing effects' and decisively contribute to project a positive external firm image and to build a good environmental and social reputation.

Nevertheless, as Frooman (1999) highlights, in order to show a positive firm image, it is a key issue to know the desires, needs and interests of firm's stakeholders, because of the effects of firm image on the firm's social reputation will depend on the social dimension considered, as well as the industry in which the firm operates. That is, the direction and strength of the corporate image–corporate reputation link is determined by the type of business activities developed in the industry, associated with its stakeholders and industry institutions. These institutional and stakeholders' pressures results are stronger if a firm's business activities have important social and environmental negative externalities, such as polluting industries like metal or paper transformation, tourism, energy generation or the like.

As Castaldo et al. (2009) assert, the ecological dimension of corporate reputation or environmental reputation occupies a key role in the today's firm's stakeholders' concerns. In this way, customers, suppliers, public administrations and so on, wonder to what extent the firm is sensitive to environmental issues, therefore being of vital importance for the firm take extremely good care of its environmental reputation.

Finally, Olmedo-Cifuentes (2011) highlights environmental and social responsibility as a key dimension of corporate reputation, and she includes as its main environmental attributes: environmental sustainability and protection, as well as the development of green products and services.

As we have proposed in previous paragraphs and mentioned in Figure 2.3, one of the key elements in determining a good corporate environmental reputation is the environmental corporate image and identity developed by the firm. In that sense, during recent years, a good green or environmental corporate image is of increasing importance, as it demonstrates the environmental indicators introduced in the measurement of corporate reputation made in the well-known corporate reputation rankings such as the *Financial Times*, *Fortune* or *Management Today*. Thus, we use in our investigation the green corporate image as a proxy and key driver of a firm's green and environmental corporate reputation.

3
Research Model

3.1 Introduction

There are many theoretical arguments for the theory that that environmental product innovation and green image play an important role as environmental management facilitators. Although, from a different perspective, both can be an important source of business opportunities and for this reason firms have started to incorporate environmental concerns into their strategic decision-making (Sharma and Vredenburg, 1998). As suggested by Karagozoglu and Lindell (2000), environmentally proactive strategies promote ecological innovation and can lead to competitive advantage.

In this way, environmental product innovation and green image, as part of environmentally proactive strategies, can help the involvement of all those stakeholders integrated in the value chain and thus generate positive returns for the firm. The idea behind this assertion is that both innovation capabilities and the firm's green image have a key function in attracting stakeholders towards more environmental commitment in their business practices, accelerating to move towards product stewardship strategies. To the extent that these sustainable business practices are able to generate income for the firm, we can argue that product stewardship activities are not only good for society but are also feasible in financial terms.

We address these issues in this section, trying to emphasize that on the one hand that both environmental product innovation and green image can improve a firm's performance separately, and on the other hand they are complementary in nature and their joint deployment can generate an indirect effect on business performance, leading to positive results for firms.

In the following section we will analyse first, the individual effects of these two concepts on firms' performance and later, their joint effect reflecting that, certainly, these two realities are interconnected.

3.2 Individual effects

3.2.1 Environmental product innovation and firms' performance

The environmental factor provides opportunities to foster innovation and develop technologies to improve efficiency. According to Hart and Milstein (2003), problems associated with industrialization like material consumption, waste and emissions represent an opportunity for companies to develop skills and capabilities in the fields of pollution prevention and ecological efficiency (Nidumolu et al., 2009).

Russo and Fouts (1995), referring to different environmental strategies as mentioned by Hart (1995), emphasize that companies that carry out pollution prevention strategies (beyond compliance with the law) focusing on environmental innovation have a resource base that enhances their ability to generate profits and also makes them able to protect themselves against future risk arising from resource depletion or the fluctuating cost of energy (Shrivastava, 1995). Dechant and Altman (1994, in Karagozoglu and Lindell, 2000), in the same vein, argue that environmental innovations enable companies to position themselves ahead of their competitors in meeting environmental regulations, which in turn helps them to protect their markets.

Therefore, the importance of incorporating environmental considerations in strategic decision-making is increasing (Sharma and Vredenburg, 1998). Thus, through environmental innovations, the firm can improve its efficiency, achieve significant cost reductions and meet the demands of those consumers especially sensitive to environmental factors.

Firms can save costs through a better use of raw materials and energy, selling the surpluses of the production process or reducing control and waste treatment cost (Murillo et al., 2008), an idea that is shared by Berrone and Gómez-Mejía (2009, in López Gamero et al., 2009), for whom proactive environmental management, characterized by innovation, can minimize waste disposal costs, reduce unnecessary steps and optimize the use of inputs among other measures. Klassen and Whybark (1999) relate environmental proactive strategies to the existence of greater opportunities for innovation and efficiency improvements. In the same line, Wagner (2005) shows that proactive environmental strategies result in the improved economic performance of the company. Rennings et al. (2006) show evidence of the relationship between environmental innovation and increased turnover of the plant, while according to Radonjic and Tominc (2007), the implementation of new and cleaner technologies lead to productivity improvements.

Therefore, with appropriate skill sets and capabilities, companies that carry out proactive environmental strategies and reduce waste emissions are able to reduce costs and increase profits (Sharma and Vredenburg, 1998; Christmann, 2000).

In addition, respect for the environment can also be a key element to meet the demands of those conscious customers who especially value the environmental performance of products, packaging and sustainable forms of business management. These customers will be willing to pay an additional price for such environmental features (Sharma et al., 1999).

Although by 2001, Hamschmidt and Dyllick argued that the market for environmentally innovative products was reduced, the development of new products that are more 'green' or sustainable has been also studied by the researchers. Despite the fact that the market for sustainable and innovative products is still in a nascent stage, the development and research about green products has been taken into consideration by a number of scholars. Among others, some of the beneficial outcomes generated by environmental product innovation are increases in sales, cost reduction and corporate image improvements (Song, in Tien et al., 2005; Chen, 2006, 2007).

Furthermore, product stewardship strategies (Hart, 1995) characterized by environmental product innovations such as product redesign or design for disassembly have been often related to better financial performance, positive image or market share improvements.

In this sense, many environmental scholars also share the same positions. Thus, Cleff and Rennings (1999), Melnyk et al. (2003) and

Pujari (2006) relate these product innovations to several indicators of financial performance. Chen et al. (2006), in turn, remark on the connection between environmental product innovation and factors such as lower costs, better management capabilities or firm growth.

More recently, Chen (2008) also found a positive and significant relationship between these innovations and the green image of the firm, while Fraj-Andrés et al. (2009) argued that, although material recycling can be expensive, it improves resource efficiency, making possible cost reductions and better innovation capabilities. In the same line, Triebswetter and Wackerbauer (2008, in López Gamero et al., 2010) argue that environmental product innovation enhances firms' competitiveness.

Given the above arguments, we can set up the following hypothesis:

H1. *The Higher the Environmental Product Innovation, the Higher the Firm's Performance.*

3.2.2 Green corporate image and firms' performance

Due to the scarcity of theoretical and empirical proposals on corporate image made from the management perspective, we will mainly use the corporate reputation literature to illustrate the strategic role of green corporate image in firms' profitability achievement.

From a general perspective, and more specifically from the management perspective, there are several contributions around the concept of corporate reputation. These contributions are coming from many different approaches: Game Theory, Stakeholders Perspective, Transaction Cost Theory or Resource-based View. Although it has been a quite common topic in the literature, there is no a general agreement about what these related concepts mean and how wide is their scope. The work of Groenland (2002: 309) tries to give some guidelines on this topic arguing that 'corporate reputation is, in its essence, a mainly emotional concept that is difficult to rationalize and to explicit.'

In this way, different approaches highlight the importance of corporate image and green reputation as well as their interrelationships. The issue can be analysed from its informative nature. From Theory Game and Information Theory approaches, businesses

situations characterized by information asymmetries, firms' images could provide important information signals acting as an effective way to avoid gaps and making predictions of future competitive and business behaviours, as well as potential reactions to firms' industry entrance and so on (Mahon, 2002). In this line, a good corporate reputation can be seen as an indicator of firms' efficiency: investment attraction, cost reduction – thorough transaction cost reduction derived from trusted relationships – incentives and positive signals to customers or attraction of talented workers (Fombrun, 1996).

From a Stakeholder Perspective (De Quevedo et al., 2007), corporate image and reputation act as links among the firm and its different main stakeholders – shareholders, managers, workers, customers, suppliers, allies, competitors and so on, and it represents an agglomeration of their assessments, perceptions and beliefs. This perspective adds value to this phenomenon in the sense of the existence of different types of firm corporate reputations and perceptions, in the way of questions like 'reputation for what?' and 'for who?'

Nevertheless, its treatment and analysis by scholars and researchers from RBV have been especially fruitful. Well-known works by Barney (1991, 1999), Grant (1991), Hall (1992) or Robert and Dowling (2002) highlight corporate reputation and corporate image as a key strategic resource/capability responsible for firms' sustained competitive advantages. This is due to its value in reinforcing differentiation product competitive strategies, as well as to its complex and social nature, its intangible character and its path dependence, firms' corporate reputation won popularity among strategy scholars as being responsible for sustained firm performance.

Thus, generally speaking, Roberts and Dowling (2002) propose a variety of potential benefits of having a favourable image and good reputation. Among these benefits we can stress the fact that a positive image and reputation is a clear indicator of the underlying quality of firms' products and services, allowing the company to fix premium prices for consumers. Furthermore, a firm with good image and reputation and that is socially and environmentally responsible may attract employees willing to work harder or for lower remuneration.

Focusing on the link between social and environmental corporate reputation, image and firms' performance, several authors propose a generally positive effect. Thus, Burke and Logsdon (1996) propose that firms will obtain a better reputation and image assessments if they develop socially responsible business activities tied to their mission, vision and corporate values, that is, linked to their corporate identities. This social reputation will allow to the firm to improve its economic performance due to public regulators and to industry actors. In this way, as Orlitzky et al. (2003) assert, instead of a direct link between social responsibility and firms' performance, there is an indirect link between firms' social and environmental activities, firms' environmental and social corporate reputation and firms' performance.

From this perspective, when firms show high degrees of social and environmental responsibility towards its stakeholders, both external and internal, this generates a positive and good corporate image among the firm's customers, shareholders, suppliers and institutional investors (Fombrun and Shanley, 1990), facilitating the access to financial resources, attracting talented workers or increasing their commitment (Orlitzky et al., 2003). This high level of firm's social responsibility builds a good firm's social reputation especially useful in those business activities where trust is a key factor in the consumer's behaviour (Castaldo et al., 2009).

In the same way, a good green corporate image, as a key driver of green or environmental corporate reputation has a decisive role in its construction and development, contributing to the firm's legitimacy towards its stakeholders (Shrivastava, 1995). As we have mentioned above, environmental activities could have 'healing effects' on a firm's reputation, by developing trust (Hart, 1995; Russo and Fouts, 1997).

The contributions from Chen (2008, 2010) highlight that environmental consciousness of consumers is more popular nowadays, and thereby, firms are enforced to enhance their environmental and social management, and in parallel, their environmental reputations. Consumers are more willing to choose and buy green products and services, and to pay high prices for products and services from firms that are environmentally responsible. In this way, firms that adopt proactive environmental strategies could have a better green image and reputation, and therefore, better profitability, constituting a virtuous business cycle.

Furthermore, Murillo et al. (2008) highlight that through the management of pollution prevention and risk, firms adapt to law and avoid future possible environmental fines and penalties, improving at the same time relationships with public administrations, financial entities, shareholders and so on. All these stakeholders increasingly take into account the firm's environmental risk in their day-to-day business activities. In this sense, the work of Gilley et al. (2000) analyses the effects that public announcements of environmental initiatives have on individual and institutional investors, which adds to their business financial analyses the environmental issues embodied in green products/services.

In summary, a good environmental corporate image in particular and environmental corporate reputation in general will contribute to trusted relationships between the firm and its key stakeholders, contributing to firms' products/services and business activities, social legitimacy and finally, contributing to firms' profitability.

Therefore, based on previous arguments we propose the positive effect of green corporate image on firm performance:

H2. *The Higher the Green Corporate Image, the Higher the Firm's Performance.*

3.3 Environmental product innovation and green corporate image: Their joint effect on firms' performance

Besides the individual effect of environmental product innovation and green image on firms' performance, if we look at the original arguments of the Natural Resource-Based View (NRBV) other complementary approaches can also be supported (Figure 3.1).

From Hart's perspective, it could be argued that environmental product innovations provide the basis from which firms can build their positive green image, which, according to our investigation,

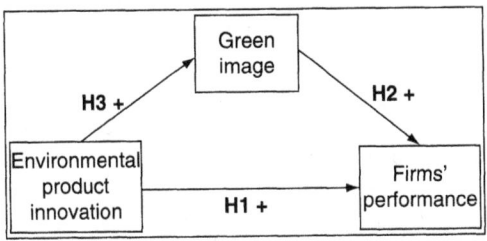

Figure 3.1 Research model: Individual and mediated effects

could also be a generator of positive results for the company. In the same vein, innovations related to the packaging design or environmentally friendly product development can be used as differentiation factors when selling products and services on the market (Murillo et al., 2008). As pointed out by Reinhardt (1998) through environmental innovations seeking product differentiation, consumers pay a higher price because of the ecological attributes of the products sold. This argument is consistent with what has been described by Fombrun and Shanley (1990), for which the fact of producing according to the criteria of social responsibility (in our case by providing an environmental argument to products) may contribute to product differentiation and enhanced corporate image and reputation.

It makes sense then, to argue that although the initial expenditures on environmental product innovation could constrain firms' capability to generate positive returns, when these innovation efforts are complemented by the generation of a positive green image, the potential economic benefits will then emerge. This means that the relationship between environmental product innovation and firm performance could be mediated by the existence of a positive green corporate image.

This line of reasoning is in the same direction as one of the main NRBV hypotheses, which states that through product stewardship strategies, firms can have access to the wide and lucrative space of environmental reputation that is not yet covered by the market (Hart, 1995; Kolk and Pinske, 2004; Orsata, 2006; Esty, 2007, in Michalising, 2010).

Furthermore, from a RBT approach, if we assume the existence of that relationship between environmental product innovation and green image, the role of the green image as a valuable, scarce and difficult to imitate resource will be reinforced. In other words, if green image would be based on the existence of certain capabilities that enable the implementation of environmental product innovations, its replication would be much more complicated and this sustainable image could be considered, in that scenario, a durable source of competitive advantage.

Summarizing, we argue that both realities are mutually supportive and can act in a complementary way. Using a similar argument to Hart's when stating that environmental proactive strategies are interconnected, and looking at the issue focusing on product stewardship strategies we can say that green corporate image mediates the relationship between environmental product innovation and firm performance.

H3. *Green Corporate Image Mediates in the Relationship Between Environmental Product Innovation and Firm Performance.*

4
Methodology

4.1 Introduction

In this chapter measures of each variable are presented in order to test previous hypotheses. First, we analyse previous empirical studies on environmental innovation with the aim of identifying current trends around this topic. Then, taking into account previous theoretical developments (see Chapter 2) and the empirical analyses regarding environmental innovation, those measures that are better adapted to the purpose of this research will be used. Then, we put forward the explanation of the questionnaire design and research resumé. Finally, focusing on the previously cited Integrated Pollution Prevention and Control (IPPC) regulations we will identify our research sample and analyse its representativeness. Consequently, one of the most polluting industries will be analysed. Particularly, the environmental behaviour (according to the variables under examination) of firms belonging to Spanish Metal Industry (production and transformation of metals), with 100 or more employees, will be studied.

4.2 Environmental innovation: Previous empirical studies

Many scholars started the research on environmental innovation at the end of the 1990s, but the empirical investigations carried out between 2000 and 2010 were the most remarkable contributions. Therefore, this topic can be considered relatively new in the management field. In general terms, we can say there is a high level of diversity among the contributions (see Table 4.1). The main methodological aspects and considerations collected by specialized scientific literature are described in the next paragraphs.

Regarding the nature of the data, cross-sectional studies are the most common, while panel studies are difficult to find because there are not enough databases on environmental innovation, making extremely difficult the realization of long-term analyses. However, Cleff and Rennings' (1999) and Horbach's (2008) works are the exception to the rule as they use Manheim's Innovation Panel (Centre for European Economic Research). This panel has a biannual nature and includes more than 1800 firms located in Germany, whose main activities are focused on industries such as mining, manufacturing, energy, construction, production and distribution, among others. Despite this the panel is not particularly focused on environmental innovation, it provides important information about new products, services and processes introduction economic success derived from that introduction and other factors that promote or prevent innovative activities within a firm.

With respect to the industries analysed in these kind of studies, most of the works are focused on manufacturing firms, such as chemicals, paper and cellulose, machinery and equipment, among others. This is due to the fact that these kind of firms realize manufacturer activities that have a direct and visible impact on the environment and, therefore, they are very concerned about the application of environmental protection measures (Bowen, 2000, in López-Gamero et al., 2010), as they must face the high cost of environmental protection (Christmann, 2000), and are directly affected by regulations (Brunnermeier and Cohen, 2003).

In regard to data collecting, as previously noted, due to the almost total absence of public environmental databases and environmental performance measures (López Gamero et al., 2010) and also because other measures (i.e. patents) are not useful to collecting information about some specific environmental innovations (i.e. environmental organizational innovations), one of the most common methods for data gathering is the questionnaire. This evidence is corroborated by prominent scholars who have stressed the adequacy of this method. According to Arundel and Kemp (2009), questionnaires are probably the best method for collecting information about results derived from environmental innovation. The utilization of questionnaires is very

Table 4.1 Previous empirical studies on environmental innovation and environmental management

Authors	Industry	Type of Investigation	Information Gathering	Geographic Area	Statistical Method	Independent Variable	Dependent Variable	Control Variable
Cleff and Rennings (1999)	Manufacturing and services	Quantitative and longitudinal	Questionnaire	Germany	Discrete choice models	Environmental innovation objectives and tools	Environmental innovation	Firm size
Klassen and Whybark (1999)	Manufacturing	Quantitative and cross-sectional	Questionnaire	USA	Multiple linear regression	Environmental technology portfolio, environmental performance	Production performance, environmental performance	Average age of teams, plant size
Christmann (2000)	Manufacturing	Quantitative and cross-sectional	Questionnaire	USA	Multiple linear regression	Best practices in environmental management, complementary assets	Cost advantage	Business unit size
Delmas (2001)	Multisectorial	Quantitative and cross-sectional	Questionnaire	USA	Structural equations	Implication degree of stakeholders for ISO 14001 standard implementation	Validity of ISO 14001 in terms of competitive advantage	Size, public price, quality management system certification, industry

Study	Sector	Method	Data source	Country	Analysis	Determinants	Dependent variable	Control variables
Brunnermeier and Cohen (2003)	Manufacturing	Quantitative and longitudinal	Secondary sources	USA	Multiple linear regression	Environmental pressures	Environmental innovation	Industry size, degree of industry concentration, capital intensity, export intensity
Melnyk et al. (2003)	Manufacturing	Quantitative and cross-sectional	Questionnaire	USA	Multiple linear regression	State of environmental management system, firm available resources, nature of corporative property	Operating income, use of environmental alternatives	None
Wagner and Schaltegger (2004)	Manufacturing	Quantitative and cross-sectional	Questionnaire	United Kingdom and Germany	Multiple linear regression	Environmental management system	Environmental competition	Firm age, industry
Ziegler and Rennings (2004)	Manufacturing	Quantitative and cross-sectional	Questionnaire	Germany	Discrete choice models	ISO 14001, EMAS, life cycle analysis, waste elimination, ISO 9001, market factors, R&D, exports, financial situation	Environmental process and product innovation	Firm age, number of installations and employees, region

Table 4.1 (Continued)

Authors	Industry	Type of Investigation	Information Gathering	Geographic Area	Statistical Method	Independent Variable	Dependent Variable	Control Variable
Del Río, P. (2005)	Manufacturing	Quantitative and cross-sectional	Questionnaire	Spain	Descriptive study. Environmental technologies adoption analysis, sources of clean technologies, barriers its adoption, and adoption impact.			
Tien et al. (2005)	Multisectorial (firms with ISO 14001 certificate)	Quantitative and cross-sectional	Questionnaire	Taiwan	Correlation models	Introduction environmental design	Competitive advantage	None
Wagner (2005)	Manufacturing	Quantitative and longitudinal	Secondary sources	European Union (Germany, Italy, Holland, United Kingdom)	Multiple linear regression	Environmental performance through inputs measures (energy and water used) and outputs measures (emissions of pollution substances)	Economic performance. ROS (Return on Sales), ROCE (Return on Owner's Capital Employed), ROE (Return on Equity)	Ratio of sales volume, debts coefficient, size, country
Chen et al. (2006)	Information and electronic technologies	Quantitative and cross-sectional	Questionnaire	Taiwan	Multiple linear regression	Green product innovation performance, green process innovation performance	Corporative competitive advantage	Firm size

Darnall (2006)	Multisectorial	Quantitative and cross-sectional	Questionnaire	USA	Discrete choice models	Regulatory pressures, market pressures, complementary resources	If firm promotes or forces to certification	Firm size and industry
Link and Naveh (2006)	Multisectorial	Quantitative and cross-sectional	Questionnaire	Israel	Multiple linear regression	Standardization	Environmental performance, firm performance	Size, time from ISO 14001 certification
Rennings et al. (2006)	Manufacturing	Quantitative and cross-sectional	Questionnaire	Germany	Discrete choice models	Maturity of environmental management system, strategic importance of EMAS, learning processes derived from EMAS, organizational scope of EMAS, measures of environmental technological and administrative innovation, environmental innovation objectives	Environmental process and product innovations; Plant economic results	None

Table 4.1 (Continued)

Authors	Industry	Type of Investigation	Information Gathering	Geographic Area	Statistical Method	Independent Variable	Dependent Variable	Control Variable
Chen (2008)	Information and electronic technologies	Quantitative and cross-sectional	Questionnaire	Taiwan	Multiple linear regression	Environmental product green innovation performance, environmental process green innovation performance, green central competition, green innovation performance	Environmental process green innovation performance, environmental product green innovation performance, green image	Size
Frondel et al. (2007)	Manufacturing	Quantitative and cross-sectional	Questionnaire	OECD (Canada, France, Germany, Hungary, Norway and USA)	Discrete choice models	Motivations (cost, image or security), environmental policy, environmental management and pressure groups	Use of End-of-pipe technologies, use of clean technologies, lack of introduction of environmental technologies	Size and sales

Radonjic and Tominc (2007)	Manufacturing (firms with ISO 14001 certificate)	Quantitative and cross-sectional	Questionnaire	Slovenia	SPSS standard analysis	Respondents reply to the following research questions: How (by what means) did the enterprises reduce environmental impacts? Which factors were improved after the introduction of new technology and to what extent? What was the origin of the new technology? How did the enterprises estimate the role of ISO 14001 certification on the adoption and updating of their technologies?	Industry, IPPC directive performance
Rehfeld et al. (2007)	Manufacturing	Quantitative and cross-sectional	Questionnaire	Germany	Discrete choice models	Existence of EMAS, elimination of waste, life cycle analysis, environmental labelling importance of regulations, R&D activities, customer satisfaction as competitive variable, exportations	Dichotomous; have you planned environmental product innovations? Which innovations have you planned to realize: environmental product or process?	Industry, firm size and firm age

Table 4.1 (Continued)

Authors	Industry	Type of Investigation	Information Gathering	Geographic Area	Statistical Method	Independent Variable	Dependent Variable	Control Variable
Wagner (2007)	Multisectorial	Quantitative and cross-sectional	Questionnaire	Germany	Discrete choice models	Industry, firm legal structure, quality management systems	Environmental product and process innovations	Firm size and firm age
Aragón-Correa et al. (2008)	Motor industry (small and medium firms)	Quantitative and cross-sectional	Questionnaire	Spain	Structural equations modelling	Innovative practices of prevention	Firm performance	Size, associations with wholesaler of sectors
Triebswetter and Wackerbauer (2008)	Multisectorial	Quantitative and cross-sectional	Questionnaire	Germany (Munich)	Case study	Environmental product innovations	Sales, market share, production cost, employment, employees qualification	Size, property and type of product
Wagner (2008)	Multisectorial	Quantitative and cross-sectional	Questionnaire	Nine European countries	Discrete choice models	Variables of environmental management, existence of quality control system, firm legal structure, environmental aspects with competitive relevance	Environmental product and process innovations	Industry, country, firm size, firm age

Rennings et al. (2008)	Manufacturing	Quantitative and cross-sectional	Questionnaire	OECD countries	Discrete choice models	Pressure groups, motivations, environmental policy, plants characteristics	Environmental management systems, adoption of a new technology	Industry
Horbach (2008)	Manufacturing and services	Quantitative and longitudinal	Secondary sources (panel data)	Germany	Discrete choice models	Increase of sales volume, environmental management, percentage of employees with high qualification, additional working time, economic situation, R&D activities, compensation/number of employees	Environmental goods and services	Firm age, industry
Darnall et al. (2008)	Manufacturing	Quantitative and cross-sectional	Questionnaire	OECD countries (Canada, Germany, Hungary and USA)	Multiple linear regression	Institutional pressures, resources and capabilities	Exhaustiveness of environmental management system, economic resources	Installations size, industry, country

Table 4.1 (Continued)

Authors	Industry	Type of Investigation	Information Gathering	Geographic Area	Statistical Method	Independent Variable	Dependent Variable	Control Variable
Iraldo et al. (2009)	Industry and services	Quantitative and cross-sectional	Questionnaire	European Union	Discrete choice models	Maturity of EMAS, environmental effectiveness, motivation towards suppliers, environmental performance	Environmental performance, market performance, innovation capability, efficiency of resources, intangible assets	Size, industry
López Gamero et al. (2009)	Multisectorial	Quantitative and cross-sectional	Questionnaire	Spain	Structural equations	Environmental proactive management, environmental performance, cost competitive advantage, differentiation competitive advantage, financial performance, early invest in environmental issues	Environmental proactive management, environmental performance, cost competitive advantage, difference competitive advantage, financial performance	Firm size

Study	Sector	Method type	Data collection	Country	Analysis	Dependent variables	Independent variables	Control variables
Wagner (2009)	Manufacturing	Quantitative and cross-sectional	Questionnaire	European Union (nine countries)	Multiple linear regression	Market share, legal form, global return, social/ environmental performance, environmental management system performance	Economic performance (measured in terms of environmental product innovation and competitive advantage)	Size, industry, country, age
Ziegler and Nogareda (2009)	Manufacturing	Quantitative and cross-sectional	Questionnaire	Germany	Discrete choice models	Environmental product and process innovation	Existence of ISO 14001 or EMAS, environmental labelling, life cycle analysis, waste elimination	Firm size, firm age, industry
Carrillo-Hermosilla et al. (2010)	Multisectorial	Qualitative	Case study	Multi-country		Eco-innovation dimensions (design, user, product/service and government) through selected cases		
López-Gamero et al. (2010)	Multisectorial	Quantitative and cross-sectional	Questionnaire	Spain	Structural equations modelling	Cost and differentiation competitive advantage, adoption of voluntary environmental steps	Financial and economic performance, environmental management, cost and differentiation competitive advantage	Size

Table 4.1 (Continued)

Authors	Industry	Type of Investigation	Information Gathering	Geographic Area	Statistical Method	Independent Variable	Dependent Variable	Control Variable
Menguc et al. (2010)	Manufacturing	Quantitative and cross-sectional	Questionnaire	New Zealand	Structural equations modelling	Relation between environmental proactive strategy (characterized by pollution prevention and environmental product and process innovation) and sales and returns gains		Industry, size and environmental dynamism
González Benito et al. (2011)	Manufacturing (plant level)	Quantitative and cross-sectional	Questionnaire	Multi country (seven OECD countries)	Moderated regression analysis	Moderator effect of environmental management system establishment on relationships between stakeholder pressures and degree of environmental performance is analysed. Direct relationship between environmental management system establishment and degree of environmental performance		Plant size, environmental organization within the firm, market scope (local, national, regional and global), firm performance

Source: Own elaboration.

common within the European Union (Smith, 2005, in Wagner, 2007) in part because the firms that patent their environmental innovations are very few (Wagner, 2007).

Among the empirical works analysed, there are also institutional surveys. Some of them utilize the Organisation for Economic Co-operation and Development (OECD) survey about environmental policy techniques (Frondel et al., 2007; Darnall et al., 2008). That survey is multisectorial and analyses the existing links among environmental policy and management, innovation investments and performance for manufacturing firms in seven OECD countries (Japan, France, Germany, Hungary, Norway, Canada and USA). In the same way, another biannual survey so-called the European Business Environmental Barometer (EBEB) is used by environmental literature (Wagner and Schaltegger, 2004; Wagner, 2007, 2008). The latest data is from 2001 and the objective was to analyse the evolution of environmental management practices within different European countries (Belgium, France, Germany, Hungary, Holland, Sweden, Switzerland and the United Kingdom). In order to gather the information, manufacturing firms were asked about their own evaluation on the environmental effects of their activities.

There is also some diversity regarding the respondents. Some studies ask the environmental manager (Rennings et al., 2006; López-Gamero et al., 2010) or the operations manager (Rehfeld et al., 2007). In our view, and taking into account that the aim of this study is related to the determination of the firm's environmental strategy, we judge it more appropriate to ask the firm's general/financial managers and environmental/innovation managers instead of focusing on single production plants. This choice is supported by strategic management literature, which asserts that although environmental management practices are carried out at plant level, according to Christmann (2000), the plants belonging to the same business unit used to apply similar practices, develop similar or linked goods, use similar productive technologies and follow the same environmental strategy.

Concerning statistical methods, there are two main trends on the topic. Most contributions utilize multiple linear regression or discrete choice models, applying sometimes structural equations modelling (within complex models in which dependent variables influence independent variables, which in turn are also affected by other independent variables). The central difference between both techniques is that discrete choice models limit the values of the dependent variable, while that restriction does not exist in multiple linear regressions.

In this sense, discrete choice models are usually applied when the purpose is to analyse whether there are or are not environmental product, process innovations or both types simultaneously (Ziegler and Rennings, 2004; Rennings et al., 2006; Rehfeld et al., 2007; Wagner, 2007, 2008), if firms promote or require an environmental certificate (Darnall, 2006), if clean environmental or control techniques are used (Frondel et al., 2007) or if there are environmental management systems (Ziegler and Nogareda, 2009).

Nevertheless, when the values of the dependent variable are not restricted, the most common practice is the utilization of multiple linear regressions. Thus, studies focused on environmental performance (Klassen and Whybark, 1999; Link and Naveh, 2006), cost advantages (Christmann, 2000), operating income (Melnyk et al., 2003), environmental competition (Wagner and Schaltegger, 2004), economic performance (Wagner, 2005, 2009), competitive advantage (Chen et al., 2006) or 'green' innovation performance (Chen, 2008) usually apply this kind of statistical method. So, as this book investigates the environmental innovation impact on a firm's performance being, therefore, the values of the dependent variable not limited, we decided to use a multiple regression technique more as the most appropriate way to carry out our empirical analysis.

Furthermore, regarding the variables in these kind of environmental studies, we can classify the independent variables into three groups: environmental performance, utilization of environmental management systems and competitive advantage derived from environmental practices. Within the first group, we can include used energy and water quantity or emission of toxic substances (Wagner, 2005), environmental technological innovations (Rennings et al., 2006; Triebswetter and Wackerbauer, 2008) or waste elimination (Rehfeld et al., 2007). In the second group, we highlight those measures referred to the development and application of environmental management systems (Melnyk et al., 2003; Wagner, 2009), the existence of the International Organization for Standardization (ISO) 14001 or the Eco-Management and Audit Scheme EMAS (Ziegler and Rennings, 2004), the maturity of the environmental management system (Rennings et al., 2006; Iraldo et al., 2009) and the utilization of environmental management methods (Frondel et al., 2007; Rehfeld et al., 2007; Horbach, 2008). And, in the third group, among environmental competitive advantage measures, we can emphasize those applied by Christmann (2000), Karagozoglu and Lindell (2000), Wagner and Schaltegger (2004) and López-Gamero et al. (2010), who employ measures about cost competitive advantage as well as differentiation of competitive advantage.

Finally, among the most cited control variables we find firm size. This can be partly explained because bigger firms have enough resources to carry out environmental innovation practices. Nevertheless, there are also contributions that stress the possibilities of small and medium firms to carry out environmental innovations (Aragón-Correa et al., 2008).

In short, according to previous empirical studies on the topic, we can highlight the following directions:

- Nowadays, due to the lack of environmental innovation data to carry out longitudinal analyses (with the exception to the rule being the Manheim's Innovation Panel, which is not focused only on environmental innovation), most of the empirical contributions are cross-sectional.
- In line with other authors who examine environmental activities within the most polluting sectors (López Gamero et al., 2010), we consider it appropriate to focus on firms belonging to the metal production and transformation industry (those firms with more than 100 employees that are included in the IPPC regulation).
- As a data gathering method, we decided to use the questionnaire because of the lack of a public database about environmental innovation (López-Gamero et al., 2010). With the aim of overcoming possible common method variance bias problems, the questions will be addressed to two different groups within a firm. Thus, the questions referred to environmental product innovation and green image will be answered by innovation/environmental managers, and those questions related to firm performance will be answered by general or financial managers.
- Finally, as our main aim is the environmental innovation impact on firm performance – without limiting values of dependent variable – we consider it appropriate to use multiple linear regressions instead of discreet choice models.

4.3 Measurement of variables

In this section environmental innovation measures are shown. Particularly, as mentioned in previous hypotheses we focus on environmental product innovation and green image as independent variables and firm performance as a dependent variable.

4.3.1 Variables development: Measurement scales

Development of variables depends on their nature. In this sense, following Diamantopoulos and Winklhofer (2001), scales can be made up of reflective (the most common ones) and formative indicators. According to Claver-Cortés et al. (2011), reflective indicators are determined by the theoretical construct, they are a reflection of it, and therefore they have a high correlation among them because all of them measure the same issue. However, formative indicators help develop the construct directly, they represent different roles of it, and therefore it is not necessary to find correlations among them, each indicator can exist independently (Podsakoff et al., 2006). Thus, conventional procedures as factor analysis and internal consistency are not appropriate for formative scales (Bollen, 1989, in Diamantopoulos and Winklhofer, 2001). Nevertheless, they make it possible to notice many concept nuances, achieving also better results. However, as other authors have pointed out (Chin, 1998, in Claver-Cortés, 2011; Diamantopoulos and Winklhofer, 2001), it is needed to assess possible multicollinearity problems for formative indicators, since several indicators of a construct could be measuring one single aspect.

Hence, taking into account the above-mentioned, our variables will be developed as reflective indicators. Then subsequent exploratory factor analyses for each one of them ('environmental product innovation', 'green image' and 'firm performance' variables) will be carried out.

4.3.2 Environmental product innovation

In regard to environmental product innovation, we have chosen a wide approach which considers the different stages in a product's life cycle. So, we will pay attention to used materials (Kammerer, 2009; Dangelico and Pujari, 2010), energy efficiency (Fiksel, 2001; Ziegler

Table 4.2 Environmental product innovation indicators

Indicator	Authors
Changes of products design in order to not use toxic compounds within production process	Based on González-Benito and González-Benito (2008)
Use of biodegradable materials for container and packaging	Dangelico and Pujari (2010)
Products design for achieving an efficient consumption of energy and water	Based on González-Benito and González-Benito (2008), Ziegler and Nogareda (2009), Dangelico and Pujari (2010)
Products design for achieving less quantity of used materials within production process	Chen et al. (2006, 2008), Dangelico and Pujari (2010)

and Nogareda, 2009; Dangelico and Pujari, 2010) and the useful life of used products (Kammerer, 2009).

In this sense, as it can be seen in Table 4.2, after the corresponding literature review on environmental innovation four indicators have been chosen.

4.3.3 Green image

In regard to green image indicators (Table 4.3), we have chosen a hybrid approach. Besides the concern about environmental issues (Chen, 2008) – more in the internal sphere of the business activity – we have also taken into account an external approach that encompasses all marketing activities that are developed to stimulate the environmental friendly attitudes and behaviours (Chamorro et al., 2009).

Table 4.3 Green image indicators

Indicator	Authors
Environmental communication	Chamorro et al. (2009), Madrigal (2000)
Environmental concern	Chen (2008)
Environmental sponsorship	Chamorro et al. (2009)
Environmental advertising	Chamorro et al. (2009), Wong et al. (1996)

4.3.4 Firm performance

Studies on environmental topics (see Table 4.4) use different approaches for linking environmental initiatives to firm performance.

Table 4.4 Studies that measure firm performance from environmental approach

Authors	Measures
Hart and Ahuja (1996)	ROA (Return on Assets), ROE (Return on Equity), ROS (Return on Sales)
Russo and Fouts (1997)	ROA
Judge and Douglas (1998)	ROI (Return on Investment), income, sales, change of market share in comparison to competitors
Sharma and Vredenburg (1998)	Several organizational capabilities
Edwards (1998)	ROCE (Return on Capital Employed), ROE
Klassen and Whybark (1999)	Production performance in comparison to competitors
Christmann (2000)	Cost advantage (position of firm cost regarding national and international competitors)
Karagozoglu and Lindell (2000)	Competitive advantage (cost, quality, reputation and international competition) and financial performance (environmental investment performance regarding returns and market share)
Delmas (2001)	Competitive advantage as degree of stakeholders implication (perception measures)
Wagner et al. (2002)	ROS, ROE and ROCE
Melnyk et al. (2003)	Operating income
Al-Tuwaijri et al. (2004)	Shares price
Wagner and Schaltegger (2004)	Environmental competition
Watson et al. (2004)	ROA, profit margin
González-Benito and González-Benito (2005)	ROA
Menguc and Ozanne (2005)	Market share, sales gain in the last two years and profit after taxes
Wagner (2005)	ROCE, ROE and ROS
Chen et al.(2006)	Corporate competitive advantage

Rennings et al.(2006)	Increase of plant economic performance, sales volume, number of employees
Link and Naveh (2006)	Firm performance (gross margin of annual profit, R&D investment, sales, sales by employee, business with foreign entities)
Cañón and Garcés (2006)	Shares price
Ann et al. (2006)	Economic and environmental impact, customer satisfaction and perceived market position
Montabon et al. (2007)	ROI, sales gain, product and process innovation
Nakao et al. (2007)	ROA, ROE, Tobin's Q and profit by share
Wahba (2008)	Tobin's Q
Iraldo et al. (2009)	Market performance (customer satisfaction and market share)
Aragón-Correa et al. (2008)	Perception measures of firm performance (invest performance and profit gain)
Darnall et al. (2008)	Economic performance (perception measures: value of installations performance and value of received orders)
Triebswetter and Wackerbauer (2008)	Competitive performance (market share, sales volume, production cost, exportations and customer base)
López Gamero et al. (2009, 2010)	Cost competitive advantage, differentiation competitive advantage, firm performance (add value gain, economic performance, financial performance)
Menguc et al. (2010)	Perception measures: firm performance (sales increase and profit gain)

Source: Own elaboration from Molina-Azorín et al. (2009).

The contributions in this field can be grouped in three different streams. In this sense, in the first group we find that many of them analyse the impact of environmental proactive and preventive strategies on firm performance (Hart and Ahuja, 1996; Russo and Fouts, 1997; Judge and Douglas, 1998; Sharma and Vredenburg, 1998; Klassen and Whybark, 1999; Christmann, 2000; Karagozoglu and Lindell, 2000; Wagner et al., 2002; Al-Tuwaijri et al., 2004; González-Benito and González-Benito, 2005; Menguc and Ozanne, 2005; Wagner, 2005; Aragón-Correa et al., 2008; López-Gamero et al., 2009, 2010; Menguc et al., 2010). In the same vein, belonging to the second group other scholars examine the relationships between environmental technological innovations and firm returns (Chen et al., 2006; Rennings et al., 2006; Montabon et al., 2007; Triebswetter and Wackerbauer, 2008). And, finally, in a third group we find other investigations which are more focused on the analysis of the role of environmental administrative innovations on firm performance (Delmas 2001; Melnyk et al., 2003; Wagner and Schaltegger, 2004; Watson et al., 2004; Ann et al., 2006; Cañón and Garcés, 2006; Link and Naveh, 2006; Darnall et al., 2008; Wahba, 2008; Iraldo et al., 2009).

Particularly, Judge and Douglas (1998) state that environmental proactive firms (those that integrate completely the environmental issues within its strategic planning) have better economic results. In order to measure these results they use perception measurement (five-point scale), asking about revenues, sales increase and changes in market share and investment performance in comparison with competitors. Klassen and Whybark (1999) link the introduction of pollution prevention technologies – characterized by offering high innovation opportunities – to improvements in production performance (using subjective and objective perceptions measures). In this case, respondents have to answer about their plant performance in comparison to competitor's plants (seven-point scale), including cost factors, product quality, speed and flexibility. Furthermore, Christmann (2000) asserts that technological innovation on pollution prevention means high firm performance, obtaining better result if firm has a great degree of complementary assets.

Taking a slightly different approach Aragón-Correa et al. (2008) use perception measures on financial performance when they analyse the influence of environmental proactive strategies on the performance of small firms. In this case, they ask about firm performance in comparison to other firms within the same industry based on investment performance and revenue gain. In a similar way, Menguc et al. (2010) utilize perceptions (five-point Likert scale) to assess firm performance. Particularly, they ask about sales increases and profit gains derived from the relationship between environmental proactive strategies and firm performance.

Empirical research focused on environmental technological innovations shows a positive relationship between green product innovation performance, green process innovation performance and firm performance (Chen et al., 2006). They use a questionnaire (five-point Likert scale) in order to measure quality, cost, image, managerial and R&D capabilities, growth, economic performance and pioneering activity. Rennings et al. (2006) present evidences on the relationship between environmental process innovation and plant returns in terms of sales volume gain and number of employees growth through telephone survey, while Triebswetter and Wackerbauer (2008) apply perception measurements in order to assess competitive performance derived from environmental innovations, including market share, sales volume, production cost, exports and customers base.

Taking a different approach, studies on environmental administrative innovation are characterized by being focused on firm performance. In this sense, Melnyk et al. (2003) analyse the relationship between environmental administrative innovation and better operating performance. The assessment of operating performance includes cost decrease, better position in the market, higher product quality, better reputation, waste minimization along the production process, better cost–benefit ratio, better new product design or development and better possibilities for international sales. They use a questionnaire with a ten-point scale from zero (totally disagree) to ten (totally agree). Wagner and Schaltegger (2004) find a positive relationship between the adoption of environmental administrative innovations and firm internal factors such as shareholders' satisfaction, board of directors' satisfaction, and employees' satisfaction.

In a similar way but obtaining different results, Link and Naveh (2006) consider that there is no relationship between environmental performance and firm performance among firms who carry out environmental administrative innovations. Firm performance measurements were collected from a database (Dan & Bradstreet database), including gross margin of annual profit, R&D investment, sales, sales by employees and business with foreign companies.

In short, as it can be seen, most of the contributions on this topic (in the management field) are focused on the discussion about whether it pays to be green or not. To do that, the great majority of authors use managers' subjective perceptions. This fact can be explained in part because managers are more inclined to provide their perceptions than to show quantitative and precise information (Aragón-Correa et al., 2008).

Table 4.5 Firm performance indicators

Indicator	Authors
ROA growth	Cohen et al. (1995), Hart and Ahuja (1996), Russo and Fouts (1997), Watson et al. (2004), González-Benito and González-Benito (2005)
ROE growth	Cohen et al. (1995), Hart and Ahuja (1996), Wagner et al. (2002), Wagner (2005)
ROCE growth	Edwards (1998), Wagner et al. (2002), Wagner (2005), Aragón-Correa and Rubio López (2007)

Table 4.6 Control variable

Indicator	Authors
Size	Cleff and Rennings (1999), Delmas (2001), Wagner and Schaltegger (2004), Ziegler and Rennings (2004), Wagner (2005), Chen et al. (2006), Darnall (2006), Link and Naveh (2006), Chen (2008), Frondel et al. (2007), Rehfeld et al. (2007), Triebswetter and Wackerbauer (2008), Wagner (2008), Iraldo et al. (2009), López Gamero et al. (2009), Wagner (2009), Ziegler and Nogareda (2009), López- Gamero et al. (2010), Menguc et al. (2010)

In this research we pay attention to financial results (ROA, ROE and ROCE), following the same line as other prominent authors within the environmental field (see Table 4.5). As the environmental innovation activities turn into better results, as a minimum, two years later, our firm performance measures are referred to the last two years (Hart and Ahuja, 1996).

4.3.5 Control variable

As a control variable (see Table 4.6), we use the most common variable considered by previous empirical studies within environmental literature: firm size (see Table 4.1).

4.4 Information sources and data gathering

Several methodological guidelines and recommendations from the previous empirical literature review (lack of database, complexity of studied phenomenon, etc.), as well as from RBV (Newbert, 2007) and Natural Resource-Based View (NRBV) theoretical frameworks, have been identified. Accordingly, our data collection and processing procedure was to the careful selection of the industry which best adapts to our research purpose. In addition, we also have to pay attention to the design of the questionnaire in order to collect data from primary sources more accurately.

As we are measuring environmental innovations, it is reasonable to think that these initiatives are, at least, determined by two main factors. On the one hand we can mention the external pressure; this way, the more polluting firms will be exposed to more pressure. On the other hand, the size; the bigger the firms are, the easier to invest in new environmental technologies.

Therefore, based on these two assumptions we decided to focus our empirical research on firms belonging to the production and transformation of the metals industry (one of the most polluting ones) with 100 or more employees (with enough resources to drive and implement environmental technologies).

Also, among the statistical considerations, we have focused on common method variance concerns that are particularly relevant, in regard to perception measures, when the same respondent replies both to independent variables and dependent variables. In order to avoid or correct common method variance (CMV), there are two general remedies, on the one hand, those remedies considered in *ex ante* research design stage and, on the other hand, those remedies implemented in *ex post* statistical analyses. 'The best way to avoid or minimize any potential CMV bias is to collect measures for different constructs from different sources' (Chang et al., 2010: 179). In short, the main recommendation is to collect dependent variables from different sources to independent variables.

Therefore, due to the fact that we use a research model in which there is an amount of independent and dependent variables, we collected data from two qualified people for each firm who replied to different sections of our questionnaire with the aim to overcome common variance bias.

The questionnaire design was carried out in several steps. First, based on previous literature and empirical studies on the environmental field, we developed a preliminary draft of the questionnaire. Second, in order to ensure the validity of the content, we consulted some experts on this subject (Govindarajan, 1988; Conca et al.,

Table 4.7 Questionnaire structure – questions directed to innovation/environmental managers

Identification data	
Environmental product innovation section	4 questions on Likert scale (1–7/1 'not consider that option', 7 'raised/proposed option is very important for us')
Green image section	4 questions on Likert scale (1–7/1 'no consider that option', 7 'raised/proposed option is very important for us')

Table 4.8 Questionnaire structure – questions directed to general or financial managers

Identification data	
Firm performance section	3 questions on Likert scale (1–7/1 'our position is worse than our competitors', 4 'our position is the same than our competitors', 7 'our position is much better than our competitors')

2004). Finally, Merka Star – a firm specializing in market research – checked the questions' phrasing.

The questionnaire was divided into four sections: the first section collects firm identification data, and the other three sections are referred to, on the one hand, independent variables of our model (environmental product innovation and green image, respectively) and, on the other hand, the dependent variable (firm performance).

Questions referred to environmental product innovation and green image were answered by innovation/environmental managers of each respondent firm, whereas questions related to firm performance were directed to general or financial managers (see Tables 4.7 and 4.8).

4.5 Sample characteristics and statistical representativeness

As it has been mentioned previously, we focused on firms belonging to industries considered in the IPPC Law16/2002. Particularly, our empirical study pays attention to production and transformation of metals industry because these kind of firms have high capacity installations and foundries and are potentially very pollutant. It is reasonable to think that firms in this industry may be aware of environmental concerns and will understand the nature of the issues related to our research.

Traditionally, production and transformation of metals industries are represented by code 24 (metallurgy; production of iron and steel products and ferroalloys) and code 25 (production of metallic products, except machinery and equipment) of NACE (NACE code is a pan-European classification system which groups organizations according to their business activities) 2009. However, in order to determine our population we have considered a wide approach, taking into account the new Industrial Emissions Directive (Directive 2010/75/EU of European Parliament and Council on November 24, 2010) of pollution integrated prevention and control, which includes industrial activities of previous IPPC regulations. In this way, using a more comprehensive analysis, we added other firms not belonging to mentioned CNAE codes, which also carry out production and transformation of metals activities. Particularly, we analysed firms belonging to the following codes (Table 4.9):

Table 4.9 Industrial activities included in the research

24	Metallurgy; production of iron and steel products and ferroalloys
25	Production of metallic products, except machinery and equipment
26	Production of computer, electronic and optical products
27	Production of electric material and equipment
28	Production of material and equipment
29	Production of motor vehicles, trailers and semitrailers
30	Production of other transport material
32	Other manufacturer industries
33	Repair and installation/system of machinery and equipment

Source: Own elaboration.

Furthermore, so that the population reflects more accurately the composition of the industry under consideration some refinement operations were driven in regard to codes 26–33. In this sense, the more relevant aspects and conclusions are the following:

- With respect to 26 and 27 CNAE codes, we have considered firms devoted to tooling, industrial lighting with aluminium, metallic cabling and wire and ferrous metals.

- Regarding 28–33 CNAE codes, firms with basic technologies related to production and transformation of metals activities were included in the population. Particularly, within self-propulsion auxiliary industry which is a global supplier of components and sub-components where basic technologies are aluminium forge, shaping of tube, welding processes, smelting and mechanized among others.

With the purpose of performing a more in-depth analysis of the industry under consideration in the next lines we will discuss some of its main features in terms of structure, turnovers or consumption figures.[1]

As can be seen in Table 4.10, turnovers have suffered important decreases owing to a great global crisis started in 2008.

Specifically, firms belonging to CNAE 25, 26 and 28 codes are the most affected industries with a reduction of 30 per cent or more in their results. Moreover, selected industries reflect a worse percentage than industry totals, with a negative 21.2 per cent compared to 17.18 per cent, respectively. This situation can be due to high relevance of selected industries within the entire industry, as they represent 34 per cent of the total with a net amount of turnover of 176,551 million Euros. Particularly, firms included in 'production of motor vehicles, trailers and semitrailers sector' (CNAE 29 code) have an important role within selected industries, since they achieve 29 per cent of selected industries' turnover, that is, 51,088 million Euros.

In addition, the number of firms in 2010 decreased 10.6 per cent within selected industries compared with 2008 (see Table 4.11). These data are in line with industry totals, which have a value of 9.1 per cent. Again, we find that selected sectors play a key role within industry total, since they involve 37 per cent of the industry total in terms of number of firms.

In regard to the activities, most firms were included in 'production of metallic products, except machinery and equipment' (CNAE 25 code), with 55.7 per cent, that is, 28,150 firms compared with 50,507 firms in all selected sectors. This percentage can be explained because firms belonging to CNAE 25 code are small- and medium-sized firms, this is a great number of firms.

Finally, paying attention to the consumptions carried out by the selected sectors firms in 2009 (see Table 4.12), electricity is the most

Table 4.10 Industries' turnover

NACE	Industries	2008	2009	2010	2008–2010 Changes %	2010 % of Total (selected sector)
24	Metallurgy; production of iron and steel products and ferroalloys	36885574	22667055	29494326	−20	16.7
25	Production of metallic products, except machinery and equipment	46190045	31925939	32433766	−29.8	18.4
26	Production of computer, electronic and optical products	9852131	6446717	6578131	−33.2	3.7
27	Production of electric material and equipment	22247345	16113349	16795522	−24.5	9.6
28	Production of material and equipment	24306590	17514049	16760499	−31	9.5
29	Production of motor vehicles, trailers and semitrailers	58084329	45984203	51088112	−12	28.9
30	Production of other transport material	15014534	12968900	12611774	−16	7.1
32	Other manufacturer industries	4407194	3743808	3956930	−10.2	2.2
33	Repair and installation/system of machinery and equipment	7184831	6586604	6831961	−4.9	3.9
Total	Selected sectors	224172573	163950624	176551021	−21.2	33.9 (industry total)
Total	Industry total	628903124	496295804	520864496	−17.18	–

Note: Turnover in thousands of Euros.
Source: INE (2012).

Table 4.11 Number of firms

NACE	Industries	2008	2009	2010	2008–10 Changes %	2010 % of Total (selected sector)
24	Metallurgy; production of iron and steel products and ferroalloys	1280	1199	1197	−6.5	2.37
25	Production of metallic products, except machinery and equipment	31997	28178	28150	−12	55.7
26	Production of computer, electronic and optical products	1753	1576	1564	−10.8	3.1
27	Production of electric material and equipment	2174	2046	1966	−9.6	3.9
28	Production of material and equipment	5734	5475	5121	−10.7	10.1
29	Production of motor vehicles, trailers and semitrailers	1810	1792	1662	−8.2	3.3
30	Production of other transport material	626	617	593	−5.3	1.2
32	Other manufacturer industries	4868	4582	4574	−6	9.1
33	Repair and installation/system of machinery and equipment	6272	6441	5680	−9.4	11.2
Total	Selected sectors	56514	51906	50507	−10.6	37.2 (industry total)
Total	Industry total	149601	136558	135966	−9.1	–

Source: INE (2012).

Table 4.12 Consumptions (year 2009)

NACE	Industries	Coal and Derivative	Oil Products	Gas	Electricity	Other Energy Consumption	Energy Consumption Total
24	Metallurgy; production of iron and steel products and ferroalloys	36346	156702	291784	1138289	23740	1646861
25	Production of metallic products, except machinery and equipment	75	70738	69232	252084	3923	396055
26	Production of computer, electronic and optical products	0	4829	1204	28625	54	34714
27	Production of electric material and equipment	248	14759	18541	112432	1295	147274
28	Production of material and equipment	16	26246	16727	85682	776	129446
29	Production of motor vehicles, trailers and semitrailers	0	19833	80016	257514	12121	369483
30	Production of other transport material	0	9861	6389	58235	1344	75829
32	Other manufacturer industries	0	3018	1624	22653	36	27330
33	Repair and installation/system of machinery and equipment	0	17004	1231	12019	890	31144
Total	Selected sectors	36685	322990	486748	1967533	44179	2858136
	% of industry total	19.5	24.3	18.6	37.5	14.4	29.5
Total	Manufacturer and extractive industry total	187808	1330578	2610574	5245388	307540	9681887

Note: Data in thousands of Euros.
Source: INE (2012).

used resource with a 37.5 per cent, followed by oil products with 24.3 per cent of manufacturer and extractive industry totals.

In this sense, we also want to highlight the relevance of the consumptions coming from firms included in 'metallurgy; production of iron and steel products and ferroalloys', and 'production of metallic products, except machinery and equipment' (CNAE 24 and 25 codes, respectively), which have the total highest energy consumption. They consume 1,646,861 and 396,055 Euros, that is, 71.5 per cent of selected sectors and 21.1 per cent of manufacturing and extractive industry totals. This result is consistent with the above considerations, since both groups of activities are traditionally the most contaminant due to their own high-capacity furnaces. These groups have, therefore, a great pollution power.

Once our population has been selected and described in their main features, it was also further refined and analysed in order to fulfil our research purposes. Particularly, those firms considered as atypical cases (outliers) were removed from the population as they had the potential to distort the descriptive statistics. Therefore, the five highest firms in terms of number of employees were removed (employees' number ranging from 3709 to 12,958). Then, our target population was finally of 733 firms with 100 or more employees.

Thus, 157 valid questionnaires were obtained (see annexes at the end), representing a response rate of 21.3 per cent, with a sampling error of +/− 5.7 per cent for a 95 per cent confidence level (see Table 4.13). After target population analysis and selection (through SABI[2] and Amadeus databases), we provide Merka Star (market investigation firm) with the population's information in order to contact them. Each survey was ten minutes long approximately, and a presentation letter explaining our proposal was sent before starting the survey (see annexes).

As we have not gathered data of all firms belonging to target population, we tested the response bias. Thus, differences between population and sample regarding size and age were examined through a t-test in order to check if our sample represents the target population (see Table 4.14). Results showed no differences based on firm's size ($p = 0.316$) and age ($p = 0.442$) for a 95 per cent confidence level because both p-values are higher than 0.05 and, therefore, null hypothesis on significant differences between sample and population means is rejected.

Table 4.13 Research facts and figures

Target population	733 firms belonging to production and transformation of metals industry
CNAE 2009	24, 25, 26, 27, 28, 29, 30, 32, 33
Firm size	100 or more employees
Geographic zone	Spain
Analysis unit	Firm
Gathering data	Telephone questionnaire
Sample size	157
Response rate	21.3%
Sampling error	+/− 5.7%
Statistical software	SPSS and AMOS 19.0
Fieldwork	March–May 2011
Respondents	– General or financial managers: they answer questions related to firm performance (see annexes – questionnaire B). – Innovation/environmental managers: they answer questions referred to environmental product innovation and green image (see annexes – questionnaire A).

Table 4.14 Sample statistical representativity

Statistic	Sample Size	Population Size	Sample Age	Population Size
Mean	27.03	28.01	274.19	247.80
Median	24	25	175	154
Mode	18	14	100	102
Standard deviation	15.78	16.10	328.88	310.30

Source: Own elaboration.

In the same way, we tested non-response bias regarding size variable (Armstrong and Overton, 1977, in Claver-Cortés et al., 2011), obtaining a p-value of 0.220 for a 95 per cent confidence level. Therefore, it can be confirmed that there is not non-response bias between respondents and non-respondents firms. On the other hand, Merka Star applied Random algorithm of Pascal's language in order to get firms for interviewing, so randomness principle of selection is met.

5
Research Results

5.1 Introduction

Now that the model variables have been identified in the previous chapter, the statistical analysis is shown in this chapter. This analysis was executed with a double objective: On the one hand, in order to analyse reliability and validity of used items and constructs and, on the other, for testing the fulfillment of the stated hypotheses.

Thus, an exploratory factor analysis was realized in order to determine the structure and dimensions of model variables through examination of factor loads and their reliability. Then, a multiple regression analysis was carried out with the aim of studying different causal relations between the identified constructors.

Nevertheless, before the statistical analysis we will, in the next section, show a global picture of the Spanish metal sector in regard to the measures of environmental product innovation and green image previously considered. This way we can analyse the current situation of the sector and later, through the empirical analysis, give some possible explanation for this evidence.

5.2 The Spanish metal sector: Overview of environmental product innovation and green image initiatives

Some considerations regarding the current situation of the sector must be addressed in order to analyse the empirical evidence later on.

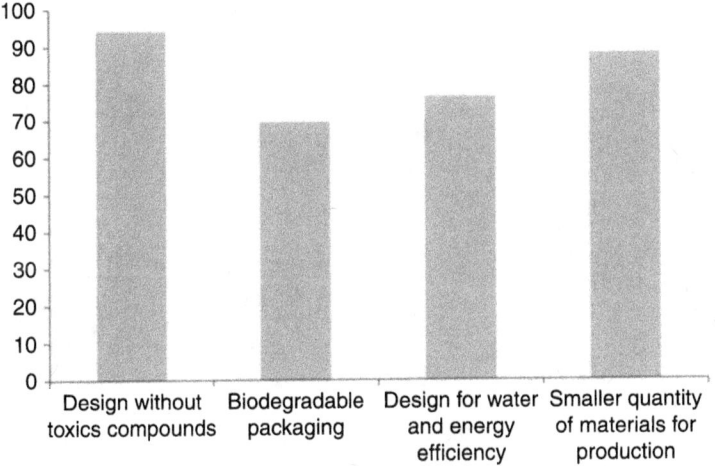

Figure 5.1 Environmental product innovation measures. Distribution in use[1]
Source: Own elaboration.

Among environmental product innovations (Figure 5.1), we can see that the most commonly used measure is the product design without toxic compounds, followed by product material reduction and the energy and water efficient design. Biodegradable packaging is in the least used measure.

These figures seem to point out that firms in the sector have two main priorities in regard to environmental product innovations. First, to avoid risk for the health of potential consumers (which could be also related to reputation risk), and second, to optimize the materials' utilization. In other words, firms are much more interested in applying those measures that have a direct impact in business activities.

In contrast, the least used measures (packaging and water and energy efficiency), although very beneficial as well, are more oriented to the public in general and produce a great part of their effects once the product is already sold.

Regarding green image (Figure 5.2), above the rest of the practices, we can stress the fact that most of the companies sees themselves as very concerned about the environmental problem. As we can see below, the situation is quite different when firms have to spend

122 Environmental Innovation and Firm Performance

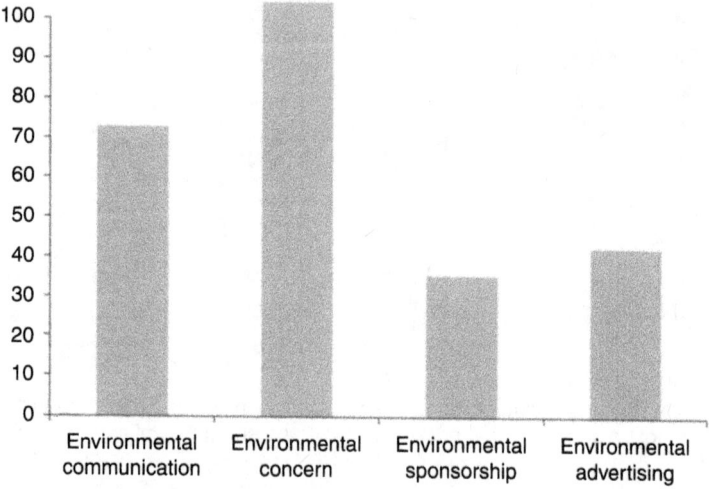

Figure 5.2 Green image measures. Distribution in use
Source: Own elaboration.

their money on this kind of initiative. Therefore, environmental sponsorship and environmental advertising are not used very often, which shows, in the same line as above, that firms are watching carefully their pockets when dealing with promotional aspects.

Furthermore, if we adopt another perspective on the analysis and attend to the degree of penetration[2] of environmental product innovation and green image initiatives, we find that, in general terms, the implementation of these two options is quite different in the business practice. In this sense, the comparison between the next two figures can shed some light on the current reality in the Spanish metal industry.

As we can see in Figure 5.3, environmental product innovations are fairly widespread among the firms in the sector, being, as previously noted, the product design for material optimization that is most frequently used.

In contrast, when analysing green image efforts (Figure 5.4), we find that these kind of measures are much less frequently used and are not widespread in the sector. All firms seem to be very concerned about environmental issues (80 per cent of the firms), but the rest of the measures are poorly implemented.

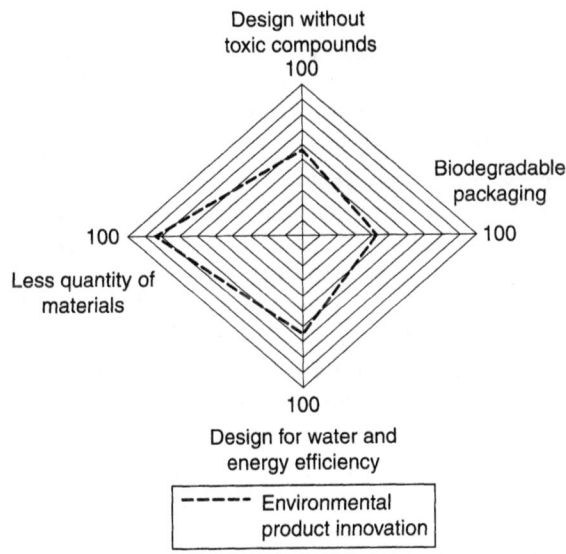

Figure 5.3 Firm profiles for environmental product innovation
Source: Own elaboration.

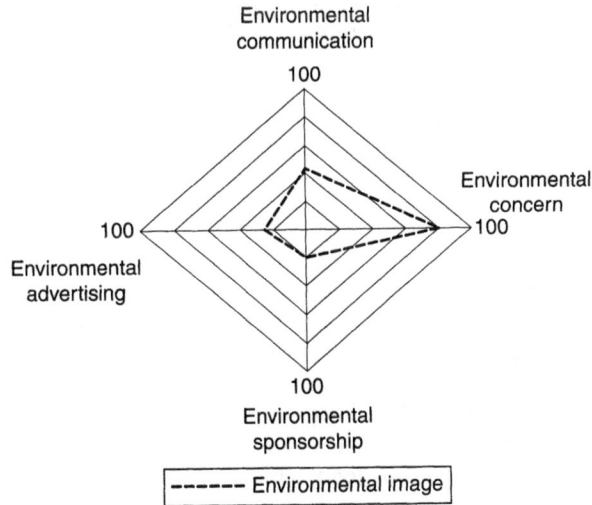

Figure 5.4 Firm profiles for green image
Source: Own elaboration.

Summarizing, if we analyse together the above considerations, two main conclusions can be highlighted:

- Firms in the sector remain very cautious with the implementation of measures that are not directly linked to their bottom lines. For example, while product design aimed to reduce the use of materials is widely used, biodegradable packaging is poorly developed. In regard to green image we can also appreciate that while most of the firms in the sector see themselves as very concerned with environmental issues, just a few of them are doing environmental advertising or sponsorship.
- Furthermore, environmental product innovation is much more widespread than green image among the Spanish metal sector. One reason for that could be that these companies (although very pollutant) are not very close to the public so they prefer an internal approach to an external one.

Once a general overview of the sector has been presented, in the next lines we will test the hypotheses previously mentioned in Chapter 3. Both the above general overview and the empirical analysis will enrich the understanding of the phenomena and will lead us to a discussion about the conclusions.

5.3 Exploratory analysis

5.3.1 Introduction

In this section, the main arguments that explain the application of exploratory factor analyses are shown, as well as some practical considerations regarding explanation of those results.

Exploratory factor analysis summarizes information contained in a data set of original variables for obtaining a smaller data set of new dimensions, with a minimum loss of information. This technique is used by studies with 100 or more cases (Hair et al., 2004). In addition, this analysis is necessary to examine the individual reliability of each item through its factor load, which is confirmed by 0.7 or more factor loads. However, some authors also consider appropriate lower loads (Chin, 1998, in Claver Cortés et al., 2011), when scores are not too low and the rest of the reliability and validity conditions are not affected with respect to that circumstance.

Table 5.1 Correlations matrix

	SizeLog	Environmental Product Innovation	Green Corporate Image	Firm Performance
SizeLog	1	0.017 (0.830)	0.108 (0.177)	0.057 (0.482)
Environmental product innovation	0.017 (0.830)	1	0.144 (0.072)	0.087 (0.278)
Green corporate image	0.108 (0.177)	0.144 (0.072)	1	0.330** (0.000)
Firm performance	0.057 (0.482)	0.087 (0.278)	0.330** (0.000)	1

Note: ** Significant at level 0.01 (bilateral).

Furthermore, this analysis allows us to study internal consistency of constructors (particularly, indicators which make them up) by means of Cronbach's alpha, which should have 0.7 or 0.6 as minimum value for exploratory investigations (Hair et al., 2004).

In addition, we must pay attention to other indexes within exploratory factor analysis: (i) matrix determinant value, which should has a value closed to 0; (ii) Kaiser-Meyer-Olkin (KMO) index with values from 0 to 1, which shows the convenience of carrying out this analysis when the value is 0.7 or more; and (iii) Barlett's test, which should have a significance level lower than 0.05 for considering the pertinence of exploratory factor analysis.

Finally, correlations between control variables and independent variables are especially checked through analysis of correlations matrix in order to examine possible multicollinearity problems (see Table 5.1). As can be seen, green image is just correlated to firm performance at a significant level of 0.000, being consistent with respect to findings found in regression analysis (see Table 5.5).

5.3.2 Environmental product innovation

As has been mentioned in the previous chapter, environmental product innovation is made up of four items, which are collected in a main dimension, and there is no evidence on multidimensional constructor within reviewed literature. Nevertheless, before that identification it is necessary to analyse the convenience of carrying out the exploratory factor analysis.

Thus, paying attention to the correlations matrix determinant value, to Bartlett's test and to the KMO measure of sampling adequacy, we can confirm that exploratory factor analysis is pertinent because the matrix determinant value is close to zero (0.200), Bartlett's test has a significance level of 0.000 and the KMO index is over 0.7 (0.786) (see Table 5.2).

Next, the 'main components' method was applied for extracting factors within factor analysis. This is the most appropriate method in order to summarize the higher amount of information (variance) in a minimum quantity of factors. In this way, we only found one factor, which explains 66.92 per cent of the total variance, being over the acceptable minimum of 60 per cent.

According to previous table data, individual reliability of each item that explain environmental innovation product is fulfilled because factor loads are over 0.7. Nevertheless, three of them have a higher relevance, being the most important of the modifications on product

Table 5.2 Main components analysis of environmental product innovation

Components Matrix	
Environmental Product Innovation Items	Factor Load
EPRODUCT (p1_1). We have modified the products design in order to not using toxic compounds within their production process	0.863
EPRODUCT (p1_2). We have modified the product packaging in order to use biodegradable materials	0.722
EPRODUCT (p1_3). We have modified the product design in order to get an efficient charge of energy and water when they were used	0.822
EPRODUCT (p1_4). We have modified the product design in order to use a lower material quantity for its production	0.857
Explained variance (%)	66.917
Accumulated variance (%)	66.917
Cronbach's alpha	0.835
Matrix correlations determinant	0.200
KMO index	0.786
Bartlett's test — Approximate chi-square	247.878
df	6
Significance	0.000

design in order to avoid toxic compounds use (p1_1), followed by use of a lower material quantity (p1_4), and to get an efficient charge of energy and water (p1_3). A lower factor load is obtained by the modifications on product packaging (p1_2).

Regarding construct reliability, Cronbach's alpha of its elements has a value of 0.835, so it is higher than the limit explained before.

5.3.3 Green image

Apart from the four items that make up green image, an exploratory factor analysis is carried out in order to identify its unidimensional condition. For that, first, it is necessary to pay attention to the correlations matrix determinant value, to Bartlett's test and to the KMO measure of sampling adequacy in order to verify its appropriateness.

As can be seen in the Table 5.3, we can confirm the convenience of realizing exploratory factor analysis because the matrix determinant value is close to zero (0.379), Bartlett's test has a significance level of 0.000 and the KMO index is close to 0.7 (0.667). Finally, one main

Table 5.3 Main component analysis of green image

Components Matrix		
Green Image Items		**Factor Load**
IMAGE (p2_1). Our company informs general public of environmental impacts and risks		0.777
IMAGE (p2_2). Our company gets worried about environmental visible aspects derived from its products and operations		0.576
IMAGE (p2_3). Our company takes part actively in the sponsorship of environmental ceremonies and events		0.784
IMAGE (p2_4). Part of our company's advertising efforts goes to stressing its environmental achievements		0.813
Explained variance (%)		55.258
Accumulated variance (%)		55.258
Cronbach's alpha		0.727
Matrix correlations determinant		0.379
KMO index		0.667
Bartlett's test	Approximate chi-square	149.378
	df	6
	Significance	0.000

factor is found, with a percentage of variance of around 55 per cent, which is appropriate within social science because it is between 50 percent and 60 per cent (Hair et al., 2004).

Particularly, advertising efforts aimed to stress environmental achievements are the most relevant item of green image (P2_4). Then, the sponsorship of environmental ceremonies and events (P2_3) and the information of environmental impacts and risks (P2_1) appear as a second and third item regarding factor load, respectively. However, the worry about environmental visible aspects has the lower factor load (P2_2).

Cronbach's alpha of 0.727 means construct reliability because is higher than 0.7, so the measurement scale is appropriate.

5.3.4 Firm performance

Firm performance is made up by three items collected in a main dimension. Before identifying its unidimensionality, the convenience of carrying out an exploratory factor analysis will be discussed.

In this sense, we pay attention to the correlations matrix determinant value, to Bartlett's test and to the KMO measure of sampling adequacy in order to confirm its pertinence. Due to the fact that the matrix determinant value is close to zero (0.086), Bartlett's test has a significance level of 0.000 and the KMO index is over 0.7 (0.767) (see Table 5.4), we can validate that pertinence. Finally, the percentage of explained variance is 88% approximately, so it is very over of 60 per cent.

The three factor loads of the main dimension (Table 5.4) have values very over 0.7, so individual reliabilities are fulfilled. Specifically, ROA has the highest value, followed by ROCE and ROE. Nevertheless, they have similar loads. Regarding construct reliability, Cronbach's alpha of 0.931 has a value very over 0.7, so the measurement scale is suitable.

5.4 Exploring the relationship between environmental product innovation, green image and firm performance

With the aim of testing hypotheses formulated within Chapter 3, causal relationships between environmental product innovation and green image and firm performance are analysed (see Table 5.5).

Table 5.4 Main component analysis of firm performance

Firm Performance Items	Factor Load
Components Matrix	
PERFORM (p2_1). With respect to competitors we are satisfied/happy with the growth of ROA in the last two years	0.940
PERFORM (p2_2). With respect to competitors we are satisfied/happy with the growth of ROE in the last two years	0.935
PERFORM (p2_3). With respect to competitors we are satisfied/happy with the growth of ROCE in the last two years	0.939
Explained variance (%)	87.977
Accumulated variance (%)	87.977
Cronbach's alpha	0.931
Matrix correlations determinant	0.086
KMO index	0.767
Bartlett's test — Approximate chi-square	378.921
df	3
Significance	0.000

Thus, using factors obtained in the exploratory factor analyses, we are going to discuss results derived from linear regression on firm performance considering firm's age and size as control variables.

As can be seen in Table 5.5, model 0 reports the regression with only the control variable. In this model, statistical F value does not show significant linear relation, so that variable does not have any effect on firm performance. Model 1 reports one of the independent variables, environmental product innovation, and model 2 reports the second one, green image. So that, when the independent variables are entered separately in the regressions, green image has significant effects on firm performance, while model 1 does not even show significant linear relation by statistical F value. Model 3 reports the full model, in which are included both independent variable, showing similar findings as model 2 on firm performance.

In the two last models, Durbin–Watson's values indicate that the residues are independent and statistical F values show a significant linear relation for proposed relationships, so there are statistical validity. Moreover, variance inflation factors (VIF) is lower than 10,

Table 5.5 Regression analysis results

	Firm Performance				Green Corporate Image	
	M0	M1	M2	M3	M4	M5
SizeLog	0.057	0.055	0.021	0.021	0.108	0.106
	(0.706)	(0.687)	(0.276)	(0.275)	(1.356)	(1.334)
Environmental product innovation		0.086		0.041		0.142*
		(1.076)		(0.527)		(1.790)
Green corporate image			0.327***	0.322***		
			(4.279)	(4.150)		
MODEL RESUME						
R	0.057	0.103	0.330	0.333	0.108	0.178
R^2	0.003	0.011	0.109	0.111	0.012	0.032
Corrected R^2	−0.003	−0.002	0.098	0.093	0.005	0.019
Typical error	1.002	1.002	0.950	0.952	0.997	0.990
Durbin–Watson	2.027	2.019	2.002	1.996	1.998	2.534
F	0.498	0.828	9.432***	6.351***	1.838	2.534*

Note: * Significant at $p < 0.10$ (t); ** significant at $p < 0.05$; and *** significant at $p < 0.01$.

so there is no multicollinearity concern. Specifically, in both models, green image has a positive and significant effect on firm performance, so hypothesis 1 is supported. However, environmental product innovation has not statistical significant effect on it, so hypothesis 2 is not supported.

Further, when green image is considered in isolation – without environmental product innovation – (M2), we obtain similar results compared to model 3, which includes the two independent variables with respect to explained variance (10.9 per cent and 11.1 per cent, respectively). Nevertheless, although green image has a positive effect and statistical significance for firm performance in two models, it has a better result within model 2.

On the other hand, considering green image as a dependent variable, model 4 reports the regression with only the control variable. Statistical F value does not show significant linear relation, so that variable does not have any effects on green image. As far as model 5 reports, environmental product innovation is an independent variable having a significant effect on green image.

In model 5, Durbin–Watson's value indicates that the residues are independent and statistical F value shows a significant linear relation for proposed relationship, so there is statistical validity. Again, variance inflation factors (VIF) is lower than 10, so there is not multicollinearity concern. Then, due to the fact that environmental product innovation influences positively on green image, hypothesis 3 is supported.

Particularly, the effect of environmental product innovation is a moderated effect ($p < 0.10$) and the explained variance has a value of 3.2 per cent. Therefore, we find that environmental product innovation has influenced firm performance through green image.

H1	The Higher Environmental Product Innovation, the Higher Firm Performance	Not supported
H2	The Higher Green Corporate Image the Higher Firm Performance	Supported
H3	Green Corporate Image Mediates in the Relationship Between Environmental Product Innovation and Firm Performance	Supported

6
Conclusions

6.1 Introduction

Now that the research proposal has been developed and the results commented upon, we now briefly outline what we think are the main contributions of this work.

In the next lines we will emphasize the importance of the Natural Resource-Based View (NRBV) as a constantly evolving theoretical framework. Although more than 15 years have passed since it was formulated, some of its aspects are still in their very early stages and many considerations have arisen. In particular, we have highlighted the role played by environmental product innovation and the green image as two interconnected realities. Both constructs have different implications to firm performance, and this fact allows us to raise other research directions that must be taken into account in the future. While the relationship between environmental product innovation and firm performance is not statistically significant, green image shows a positive impact on firm performance.

Nevertheless, we think that the most relevant evidence has to do with the direct effect of environmental product innovation on green image. As previously noted, the connection between environmental product innovation and firm performance is only achieved through green image, which shows the mediating role of the latter.

In addition, in this book, we have formulated a specific conceptual delimitation of environmental innovations. Using a novel concept of environmental innovation, we will emphasize its socio-economic character and also describe the general situation in which the Spanish metal sector, regarding environmental innovation and green image, is currently immersed.

On the basis of these analyses in the last part of the book we will look carefully at the empirical evidence that supports the above arguments, assessing, at the same time, the implications for future development of the field. The book ends with the final reflection about the limitations and future research directions.

6.2 The evolving nature of the Natural Resource Based View

At the theoretical level, the NRBV, as an extension of the well-known RBV, has shown its relevance and up-to-date postulates. Thus, every day seems more evident that the constraints imposed by the natural environment are strongly undermining the firm's capabilities to generate wealth and prosperity or even conditioning its survival.

According to this approach, firms must pass through several stages in order to meet the challenge of the natural environment. To compete in the markets of the future, firms must develop environmental proactive strategies that go from pollution prevention to product stewardship and ultimately to sustainable development. This way, it will be possible to reach cost advantages at the very first stage, to subsequently pre-empt competitors through the preferential access to certain resources or by setting up the rules or the standards better adapted to firm capabilities. All to achieve, ultimately, the best possible position in developing markets using a long-term and environmentally conscious strategy.

Nevertheless, without subtracting importance from the above considerations, the fact is the development of the NRBV has been uneven and limited (Michalising and Stinchfield, 2010). While research in pollution prevention have taken a great part of environmental scholars' efforts, product stewardship, one of the main arguments of this book, has been almost ignored in the environmental management literature (Hart and Dowell, 2010), which demonstrates that this theoretical framework is still to be developed. Furthermore, it is also argued that in order to empirically test Natural Vision hypotheses and Hart's strategies, both the establishment of more explicit relationships (Buysse and Verbeke, 2003) and the understanding of the contingencies affecting the relationship environmental performance-financial performance are needed (Berchicci and King, 2007, in Hart and Dowell, 2010).

Thus, taking into account the current development of the NRBV, one of the main contributions of this book lies in the empirical assessment of some of the aspects that have received less attention within this approach. Therefore, some particular manifestations of product stewardship strategies and green image initiatives will be analysed, establishing at the same time both individual and joint relationships between these two concepts and firm performance and giving as well some orientations about green competences (Buysse and Verbeke, 2003) that support both of them.

With the analysis of environmental product innovations, some aspects of product stewardship strategic capability are assessed, showing a very innovative and not frequently used environmental product innovation-firm performance approach.

In the same vein and in order to complete the analysis, a firm's green corporate image is also evaluated. While environmental product innovation is using an internal approach, the projected green image uses an external orientation that reflects outside the internal efforts carried out by the firm. This external approach will, ultimately, have the purpose of involving the stakeholders in order to achieve the already mentioned sustainable development strategy.

Nevertheless, this book is not only limited to the analysis of NRBV strategies but also tries to offer new theoretical reflections and prospects in the environmental field. In this sense, with the main aim of connecting environmental innovation with conventional literature on innovation, this work presents an objective-based classification of innovations, showing that these can have different manifestations depending on whether the objective pursued is economic, social or socio-economic in nature. Thus, particularly, environmental innovations are socio-economic innovations as they enable new business opportunities as well as environmental sustainability for future generations.

In the next section, we will show the implications that for the future development of the theoretical framework have the empirical results outlined above.

6.3 Implications

Once we have emphasized the relevance of the NRBV, it is important to have a global picture of the work, so we can analyse to what extent the conclusions obtained after the empirical treatment support the evolution of the theoretical framework or help the development of new perspectives and approaches in the field of environmental strategies.

Therefore, in this section we will show what we think are the main implications derived from the relationships raised in the previous chapters of the book. These analyse the role of environmental product innovation in the development of the NRBV and at the same time show the external projection, through the analysis of the green image, of this theoretical framework.

Accordingly, we must highlight the role played by environmental product innovations. Through them, the transition from traditional control and prevention approaches to the almost unexplored field of product stewardship (Hart and Dowell, 2010) takes place. A number of different aspects are analysed in the casual relationships under examination, and the role played by environmental product design in general and other measures like toxic compounds elimination and the use of biodegradable, efficient and recyclable materials is discussed. Therefore, with these relationships we try to determine to what extent certain aspects of product stewardship are useful to reduce a firm's cost and ultimately generate positive business results.

Although with the empirical evidence we cannot state clearly that environmental product innovation measures result in better firm performance, is it also true that these innovations seem to be connected to the green image of the firm and also, indirectly, with the firm's performance as well. Furthermore, our results indicate that, if firms are unable to configure the link between environmental product innovation and green image, the innovation efforts will not come true. In other words, firms' expenditures in green product innovation need to have much more visibility to the public to offset, through green image, the initial cost derived from innovation.

As suggested in the previous hypotheses, this approach is in line with the NRBV, which stated that besides the benefits in terms of efficiency and cost, environmental proactive strategies can also pay off in terms of image and reputation.

Furthermore, this work also presents some RBT implications. The effect of environmental product innovation on green image could be interpreted as an additional demonstration of the relevance of firms' intangible assets. In this sense, green corporate image emerges as a valuable, scarce and difficult-to-imitate resource, and as a crucial element in the success of environmental product innovations.

In short, our conclusions support some of the main arguments of the NRBV but also give a glimpse of other aspects that are still in a nascent or preliminary stage. Besides stressing the importance of concepts like green image or green reputation, this book intends to contribute to the development of our theoretical framework. The empirical analysis of product stewardship measures, as previously noted, provides another starting point in the development of the NRBV, as these strategies are much wider than we have shown in this book and cover the entire value chain integrating all the stakeholders and leading to the creation of green or sustainable reputation in the firms that implement them.

Finally, is it worth emphasizing that empirical evidence seems to indicate that firms in the Spanish metal sector prefer to devote their efforts to the development of their environmental product innovations rather than to the creation of their sustainable image. This is quite remarkable since the returns obtained from green image initiatives are much more important in financial terms. Therefore, it seems like the firms in this industry are leaving aside image considerations and are more technologically centred, evidence that is far from being the most profitable solution. In conclusion, in order to corroborate this first impression, many additional efforts must be done in the future.

6.4 Limitations and future research lines

Without belittling the above considerations, this work has also the limitations which are detailed in the next paragraphs.

First, regarding the environmental innovation concept we have developed in this book, we want to stress the fact that, although we think that the environmental innovation must be related to the conventional literature on innovation, having also a socio-economic character, other approaches to the term can be equally valid if properly formulated. This is the case of the contributions, out of the scope of this book, that place the green intellectual capital at the genesis of the environmental innovation concept.

Second, we want to highlight also how difficult is to find measures that are appropriate for the purpose intended in this research. Thus, we have chosen to use the questionnaire that, although is based on subjective perceptions, is widely accepted in this kind of studies. As Arundel and Kemp (2009) and many environmental scholars have stated, the questionnaire is probably the best method to gather information in order to assess environmental innovation results. Nevertheless, it would be very useful for the future, as a way to corroborate the validity of the results, to compare these perceptions with more objective measures derived from secondary sources. This could help the researchers to assess to what extent the respondents' perceptions reflect the real economic activity.

As the third and maybe the most important limitation, it is worth highlighting that the lack of environmental databases of public access (López-Gamero et al., 2010) conditions in large measure this kind of studies. Thus, today, most environmental innovation studies are cross-sectional, which does not allow the realization of greater time scope analysis that certainly would improve the quality and explanatory power of the analysis.

And last, other limitation has to do with the generalization of the results. These, which are referred to businesses with more than 100 employees, are not applicable to smaller firms, which represent the majority of Spanish industry and whose environmental commitment must be promoted and analysed in order to change the current economic model to one more sustainable both in economic and environmental terms. In the same vein, it would be highly interesting to expand the scope of this research to other industries and services as well as other countries to carry out more clarifying comparative analysis.

Future research lines on this topic would be mainly directed to address the previously mentioned limitations. Therefore, we think that future contributions aimed to improve both the measures and the generalization of the results (e.g. other industries or countries) would greatly enrich the knowledge about environmental innovations and environmental strategies in general.

It would also be of great interest to have access to environmental innovation databases which certainly would shed some light about the evolution of the environmental commitment among firms and also would lead to more solid results.

Ultimately, we have to say that the empirical results give us a glimpse of what will happen with some future research lines that have been only partially addressed in this book and that would contribute to the development of the NRBV. For example, the emergence of environmental management systems provides other promising research lines. The role of the environmental management systems and the stakeholders as moderating factors in the environmental product innovation-firm performance relationship is still to be clarified and could have a decisive influence on the generation of positive returns for companies. Therefore, the supporting role of environmental management systems must be clarified, given that these management tools could boost the incomes derived from environmental product innovations and also play a role both as innovation and coordination facilitators.

As previously noted, product stewardship strategies, as the second stage of the NRBV, must be addressed from a comprehensive perspective in order to introduce in the environmental discourse all and each one of its facets. Topics like the stakeholders' involvement (González-Benito et al., 2011) in green product design and the environmental sustainability of value chain' activities constitute a novel and important research avenue for the future and would help us to check to what extent firms are moving towards the environmental sustainability of its operations. Although in this book we have considered some product design features that connect product stewardship strategies with the stakeholders, it is also true that a more comprehensive analysis could clarify to what extent these can influence the development and ultimately, the success of environmental product innovations. Thus, for example, issues related to the connections between environmental product innovations and suppliers integration are critical to the generation of better business results.

As Aragón-Correa and Sharma (2003) have pointed out, the inclusion of exogenous factors in the analysis and its influence in the development of environmentally proactive strategies could be an interesting research avenue for the future and also, according to our results, a profitable strategy for firms within this sector.

Annexes

ANNEX 1. Carta de Presentación y Cuestionario (Spanish)

Estimado Sr./Sra.,

Nos ponemos en contacto con usted para solicitar su colaboración en el estudio que, desde la Universidad Complutense de Madrid y con el apoyo del Ministerio de Ciencia e Innovación, estamos llevando a cabo.

El objetivo del estudio es conocer, desde su experiencia profesional, el impacto de las distintas prácticas de innovación medioambiental en los resultados y la ventaja competitiva de las empresas. Los resultados obtenidos podrían ser de gran valor tanto para la comunidad científica como para las Administraciones Públicas, al constituir una importante referencia de cara a la futura elaboración de políticas orientadas a alcanzar la sostenibilidad económica y medioambiental de la actividad empresarial en nuestro país.

Con tal propósito, solicitamos su colaboración y nos dirigimos a su empresa como importante referente del sector al que pertenece.

Agradecemos de antemano la atención prestada y si fuera posible su predisposición en el estudio mencionado.

Le saluda atentamente, quedando a su disposición

Javier Amores Salvadó
Profesor Organización de Empresas
Universidad Complutense de Madrid.
Campus de Somosaguas. 28223 Madrid. jamores@ccee.ucm.es

ANNEX 1. Questionnaire and Cover Letter

Dear Mr./Mrs.,

We are contacting you to ask for a possible collaboration in a research project which is being developing in the Complutense University of Madrid, and supported by the Spanish Ministry of Science and Innovation.

The aim of this study is to know, based on your professional experience, the impact of different environmental innovation practices on firm's competitive advantage and performance. The obtained results will be of great value to the scientific community, industry practitioners, and the Public Administrations, because they could be important references for future public policy developments in order to achieve firm's economic and environmental sustainability in our country.

To this end, we are contacting you and your firm as industry's key actors requiring your appreciated collaboration.

Thanks in advance for your time and attention, and if it would be possible, your willingness to cooperate in our study.

Yours sincerely,

Javier Amores Salvadó
Assistan Professor of Business Administration
Universidad Complutense de Madrid.
Campus de Somosaguas. 28223 Madrid. jamores@ccee.ucm.es

CUESTIONARIO. PARTE A (In Spanish)

Datos de identificación de la empresa:

- Nombre de la empresa
- Nombre y cargo de la persona encuestada

Usando una escala de 1 a 7, donde 1 significa 'ni siquiera hemos considerado esa alternativa', y 7 'la alternativa planteada es de vital importancia para nosotros', dígame el grado de implantación en su empresa de cada una de las prácticas que le enumero a continuación en relación con la **Innovación Medioambiental de Producto**

Hemos modificado el diseño de los productos para que no se utilicen compuestos tóxicos en su elaboración.	1 2 3 4 5 6 7
Hemos modificado el envase y embalaje de los productos de forma que utilicen materiales biodegradables.	1 2 3 4 5 6 7
Hemos modificado el diseño de los productos para que al ser utilizados sean eficientes en el consumo de energía y agua.	1 2 3 4 5 6 7
Hemos modificado el diseño de los productos de forma que sea necesaria una menor cantidad de material para su elaboración.	1 2 3 4 5 6 7

Usando la misma escala anterior, dígame el grado de implantación en su empresa de cada una de las prácticas que le enumero a continuación en relación con la **Imagen Verde**.

Nuestra compañía comunica los impactos y riesgos medioambientales al público en general.	1 2 3 4 5 6 7
Nuestra compañía muestra preocupación por los aspectos medioambientales visibles derivados de sus productos y operaciones.	1 2 3 4 5 6 7
Nuestra empresa participa activamente en el patrocinio de actos y eventos medioambientales.	1 2 3 4 5 6 7
Nuestra empresa destina parte de sus esfuerzos publicitarios a destacar sus logros medioambientales.	1 2 3 4 5 6 7

CUESTIONARIO. PARTE B (Spanish)

Datos de identificación de la empresa:

- Nombre de la empresa
- Nombre y cargo de la persona encuestada

Resultados Empresariales

Ya para finalizar, dígame, por favor, la posición de la empresa respecto de los competidores siendo 1 'nuestra posición es peor que la de la competencia', 4 'estamos a la par que nuestros competidores' y 7 'superamos claramente a nuestros competidores'.

En relación a nuestros competidores, estamos satisfechos con el crecimiento mostrado por la rentabilidad económica o ROA en los dos últimos años.	1 2 3 4 5 6 7
En relación a nuestros competidores, estamos satisfechos con el crecimiento mostrado por la rentabilidad financiera o ROE en los dos últimos años.	1 2 3 4 5 6 7
En relación a nuestros competidores, estamos satisfechos con el crecimiento mostrado por la rentabilidad del capital invertido o ROCE en los dos últimos años.	1 2 3 4 5 6 7

QUESTIONNAIRE, PART 'A'

Company's data:

- Company name:
- Respondent's name, surname and firm's title/position:

Using a perceptual 1–7 Likert scale, where 1 means 'we have not even considered that option' and 7 'the alternative proposed is of vital importance to us', please tell us the degree of real implementation in your firm for each of the practices listed below considering **Environmental Product Innovation**.

We have modified the product's design in order to avoid toxic compounds in its elaboration/production.	1 2 3 4 5 6 7
We have modified the product's packaging to use biodegradable materials and compounds.	1 2 3 4 5 6 7
We have modified the product's design to make its use in terms of water and energy consumption more efficient.	1 2 3 4 5 6 7
We have modified the product's design to reduce the quantity of material required in its elaboration/production.	1 2 3 4 5 6 7

Using the same perceptual 1–7 Likert scale, please tell us the degree of real implementation in your firm for each of the practices listed below considering your **Green Corporate Image**.

Our company communicates environmental risks and impact to society in general.	1 2 3 4 5 6 7
Our company is concerned about visible environmental aspects arising from its products, services and business activities.	1 2 3 4 5 6 7
Our company is actively involved in sponsoring environmental events and meetings.	1 2 3 4 5 6 7
Our company devotes part of its marketing efforts to highlight its environmental achievements.	1 2 3 4 5 6 7

QUESTIONNAIRE, PART 'B'

Company's data:

- Company name:
- Respondent's name, surname and firm's title/position:

Firm performance

Using a perceptual 1–7 Likert scale, please tell us your company's competitive position in relation to your competitors, where 1 means 'our firm's position is worse than our competitors'', 4 means 'our firm's position is on a par with our competitors'' and 7 means 'our firm's position clearly surpasses our competitors''.

In relation to our main competitors, we are satisfied with the firm's economic performance or ROA growth during the last two years.	1 2 3 4 5 6 7
In relation to our main competitors, we are satisfied with the firm's financial performance or ROE growth during the last two years.	1 2 3 4 5 6 7
In relation to our main competitors, we are satisfied with the firm's capital employed performance or ROCE growth during the last two years.	1 2 3 4 5 6 7

ANNEX 2. Firms Included in the Research Sample

1	ABANTIA TICSA SA
2	ACERIA DE ALAVA SA
3	ACEROS PARA LA CONSTRUCCION SA
4	ACG CENTRO BAHIA S.L.
5	ALBATROS ALCAZAR SOCIEDAD ANONIMA.
6	ALCOA TRANSFORMACION DE PRODUCTOS SL
7	ALUMINIO CANDIDO SOCIEDAD ANONIMA.
8	AMADEO MARTI CARBONELL SA
9	ANDREU BARBERA SL
10	ARALUCE SA
11	ARC DISTRIBUCION ARTE PARA EL HOGAR IBERICA SL.
12	ARMADURAS DEL SUR S.L.
13	ARTEIXO TELECOM SA
14	ASIENTOS DE GALICIA S.L.
15	ASTURFEITO SA
16	ASTURIANA DE ZINC S.A.
17	AZKOYEN INDUSTRIAL SA
18	BELLOTA HERRAMIENTAS SA
19	BEZARES SA
20	BORGWARNER EMISSIONS SYSTEMS SPAIN SL
21	BOST MACHINE TOOLS COMPANY SA
22	BROSE, S.A.
23	BRUSS JUNTAS TECNICAS S.L. SOCIEDAD EN COMANDITA.
24	CABLES RCT SA
25	CABLES Y ALAMBRES ESPECIALES SA
26	CAMETAL SA
27	CANTAREY REINOSA S.A.
28	CARPINTERIA DE PVC EUROVENT SL
29	CELULOSA FABRIL SA
30	CIE LEGAZPI S.A.
31	CLEM SA

(Continued)

32	COHIDRANE SUR S.L.
33	COMPONENTES AERONAUTICOS COASA S.A.
34	CONDESA FABRIL SA
35	CONSTRUCCIONES FERROVIARIAS DE MADRID S.L.
36	DALPHI METAL ESPANA SA
37	DEUTZ DITER SA
38	DIMETRONIC SA
39	E D S INGENIERIA Y MONTAJES SA
40	EGANA 2 SOCIEDAD LIMITADA.
41	EKIN S.COOP.
42	ELECTRO CRISOL METAL SA
43	EQUIPOS NUCLEARES SA
44	ESTAMPACIONES METALICAS EPILA S.A.
45	EUROCIR SA
46	EUROFREN BRAKES SL
47	EXTRUSIONES DE TOLEDO SA
48	FABRICACION Y REPARACION DE BUSES SL
49	FAGOR EDERLAN SOCIEDAD COOPERATIVA LIMITADA
50	FEDERAL-MOGUL FRICTION PRODUCTS SA
51	FELGUERA CALDERERIA PESADA SA
52	FLEXOL ESPANA SL
53	FUNDICION NODULAR SA
54	FUNDICIONES DE ODENA SA
55	FUNDICIONES DEL ESTANDA SA
56	GADIR SOLAR S.A.
57	GADITANA DE CHORRO Y LIMPIEZA SL
58	GESTAMP PALENCIA SA
59	GESTAMP TOLEDO SL
60	GRUPO ANTOLIN DAPSA SA
61	GRUPO ANTOLIN RYA SA
62	GRUPO ARMARIOS PERSIANA S.A.
63	HAYES LEMMERZ MANRESA SL
64	HIAB CRANES SL.
65	HORTA-COSLADA CONSTRUCCIONES METALICAS SL.
66	HUSQVARNA ESPANA S.A.
67	IBP ATCOSA S.L.
68	INASA FOIL SA
69	INDUSTRIAL DE ELEVACION SA
70	INDUSTRIAL GRADHERMETIC SAE
71	INDUSTRIAS DUERO SA
72	INDUSTRIAS EMILIO DIAZ SA
73	INDUSTRIAS GUERRA SA
74	INDUSTRIAS METALICAS ANRO S.L.
75	INDUSTRIAS RAMON SOLER SA
76	INDUSTRIAS SALUDES, S.A.

77	INDUSTRIAS TEIXIDO SA
78	INFRICO SL
79	INOXFIL SA
80	INSONORIZANTES PELZER SA
81	INTEC-AIR SL
82	ITW METAL FASTENERS S.L.
83	KLEIN IBERICA SA
84	KNIPPING ESPAÑA SA
85	LAMP SA
86	LIDERKIT SL
87	LUXE PERFIL SL
88	M P PRODUCTIVIDAD S.A.
89	M TORRES DISENOS INDUSTRIALES SA
90	MAC PUAR AERONAUTICA SOCIEDAD LIMITADA.
91	MANN-HUMMEL IBERICA SA.
92	MATRICERIA Y ESTAMPACION F SEGURA SL
93	MAVE AERONAUTICA S.L.
94	MB ABRERA S.A.
95	MB LEVANTE S.L.
96	MECANIZACIONES Y FABRICACIONES SA
97	MECANIZADOS IKETZ SL
98	MEGASA SIDERURGICA SL
99	MENDIGUREN Y ZARRAUA SA
100	METALBAGES P51 SL
101	METALCOMPONENTES MB NAVARRA S.A.
102	METALGRAFICA GALLEGA S.A.
103	METALURGICA CERRAJERA DE MONDRAGON SA
104	MIERES TUBOS SOCIEDAD LIMITADA.
105	MIGUELEZ SL
106	MITSUBISHI MATERIALS ESPANA SA
107	MODULAR LOGISTICA VALENCIANA SL.
108	MONTRASA MAESSA ASTURIAS S.L.
109	NONCALES SL
110	NOVACERO S.A.L.
111	OERLIKON SOLDADURA SA
112	PANEL FIJACIONES SOCIEDAD COOPERATIVA.
113	POLTANK SA
114	PORTALAMPARAS Y ACCESORIOS SOLERA SA
115	PRODUCCIONES MITJAVILA SOCIEDAD ANONIMA.
116	PRODUCTOS TUBULARES SA
117	PROTECCIONES Y LACADOS SL
118	QUALITY ESPRESSO SA
119	RELECO SA
120	REMOSUR S.L.

(Continued)

121	REUNION INDUSTRIAL SL
122	ROLDAN SA
123	RONAL IBERICA SA
124	ROTULOS AYLLON SA
125	S.A. DE SOLDADURA Y METALIZACION ATOMIZADA
126	SA NJO DE ESTAMPACIONES
127	SAINT-GOBAIN PAM ESPANA SA
128	SAPA PROFILES NAVARRA S.L.
129	SENALIZACIONES VILLAR SA
130	SILICIO SOLAR S.A.
131	SISTEMAS MECANICOS AVANZADOS SL
132	SISTEMAS TECNICOS DEL ACCESORIO Y COMPONENTES S.L.
133	SMR AUTOMOTIVE SYSTEMS SPAIN SOCIEDAD ANONIMA
134	SOCIEDAD ESPANOLA DE ELECTROMEDICINA Y CALIDAD S.A.
135	SOME SA
136	TALLERES DE ESCORIAZA S.A.
137	TALLERES FELIPE VERDES SA
138	TALLERES ZITRON SA
139	TAPON CORONA IBERICA SA
140	TARABUSI SA
141	TECNOCONFORT SA
142	TECNOLOGIA Y MONTAJE SA
143	TECNOTRANS BONFIGLIOLI SA
144	TELERGON SA
145	TINCASUR SUR S.L.
146	TMD FRICTION ESPANA S.L.
147	TRELLEBORG AUTOMOTIVE SPAIN SA
148	TRELLEBORG IZARRA, S.A.
149	TRENZAS Y CABLES DE ACERO PSC S.L.
150	TRISTONE FLOWTECH SPAIN SA
151	TROQUENOR SA
152	T-SOLAR GLOBAL S.A.
153	TUBOS DEL MEDITERRANEO SA
154	TUBOS REUNIDOS SA
155	VICENTE TORNS SA
156	ZANINI EPILA S.L.
157	ZAYER S.A.

Notes

1 Theoretical Framework

1. *Source*: United Nations Documentation, http://www.un.org/depts/dhl/resguide/specenv.htm.
2. *Source*: United Nations Documentation, http://www.un-documents.net/ocf-ov.htm.
3. Hart and Dowell call them either strategic capabilities or stages within the environmental proactive strategies.
4. Hart (1995: 995) gives interesting examples referring to the strategic movements of firms like Wal-Mart, Du Pont and BMW.

2 The Nature of Environmental Product Innovation and Green Image

1. Well-documented examples of these eco-innovations can be found in Hellström (2007).
2. The concept of clean technology used in this work is in line with Rennings et al. (2006) referred to the environmental impact during the process. By contrast, Hart (1995) and Hart and Dowell (2010) relate clean technologies to radical innovations.

4 Methodology

1. 2010 data.
2. SABI stands for "Sistema Anual de Balances Ibéricos".

5 Research Results

1. Number of firms adopting mentioned measures.
2. Percentage of firms using the analysed measures.

References

Acedo, F. J., Barroso, C. and y Galán, J. L. (2006). 'The Resource-Based Theory: Dissemination and Main Trends'. *Strategic Management Journal*, 27, 621–636.
AENOR (2010). http://www.aenor.es/documentos/certificacion/folletos/w_439_EMAS.pdf.
AENOR (2010). http://www.aenor.es/documentos/certificacion/folletos/w_436_ISO14001.pdf.
Al-Tuwaijri, S. A., Christensen, T. E and Hughes, K. E. (2004). 'The Relations Among Environmental Disclosure, Environmental Performance, and Economic Performance: A Simultaneous Equations Approach'. *Accounting, Organizations and Society*, 29, 447–471.
Amit, R. and Shoemaker, P. (1993). 'Strategic Assets and Organizational Rent'. *Strategic Management Journal*, 14, 33–46.
Andersen, M. (2008). Eco-Innovation – Towards a Taxonomy and a Theory. 25th Conference on Entrepreneurship and Innovation – Organizations, Institutions, Systems and Regions. June 17, Copenhagen.
Ann, G., Zailani, S. and Wahid, N. (2006). 'A Study of the Impact of Environmental Management System (EMS) Certification Towards Firms' Performance in Malaysia'. *Management of Environmental Quality*, 17, 73–93.
Aragón-Correa, J. A., Hurtado-Torres, N., Sharma, S. and García-Morales, V. J. (2008). 'Environmental Strategy and Performance in Small Firms: A Resource-Based Perspective'. *Journal of Environmental Management*, 86, 88–103.
Aragón-Correa, J. A. and Rubio-López, E. (2007). 'Proactive Corporate Environmental Strategies: Myths and Misunderstandings'. *Long Range Planning*, 40, 357–381.
Aragón-Correa, J. A. and Sharma, S. (2003). 'A Contingent Resource-Based View of Proactive Corporate Environmental Strategy'. *Academy of Management Review*, 28, 71–88.
Arundel, A. and Kemp, R. (2009) *Measuring Eco-Innovation*. United Nations University, Working Paper Series.
Barnett, M., Jermier, J. and Lafferty, B. (2006). 'Corporate Reputation. The Defitional Landscape'. *Corporate Reputation Review*, 9, 25–38.
Barney, J. (1986). 'Strategic Factor Markets: Expectations, Luck, and Business Strategy'. *Management Science*, 32, 1231–1241.
Barney, J. (1991). 'Firm Resources and Sustained Competitive Advantage'. *Journal of Management*, 17, 99–120.
Barney, J. (1999). 'How a Firm's Capabilities Affect Boundary Decisions'. *Sloan Management Review*, Spring, 137–145.
Barney, J. (2001). 'Resource-Based Theories of Competitive Advantage: A Ten Year Retrospective on the Resource-Based View'. *Journal of Management*, 27, 643–650.

Barney, J., Ketchen, D. and Wright, M. (2011). 'The Future of Resource-Based Theory: Revitalization or Decline?' *Journal of Management*, 37, 1–15.
Barreto, I. (2010). 'Dynamic Capabilities: A Review of Past Research and an Agenda for the Future'. *Journal of Management*, 36, 256–280.
Bontis, N. (1998). 'Intellectual Capital: An Exploratory Study That Develops Measures and Models'. *Management Decision*, 36, 63–76.
Brunnermeier, S. B. and Cohen, M. A. (2003). 'Determinants of Environmental Innovation in US Manufacturing Industries'. *Journal of Environmental Economics and Management*, 45, 278–293.
Buysse, K. and Verbeke, A. (2003). 'Proactive Environmental Strategies: A Stakeholder Management Perspective'. *Strategic Management Journal*, 24, 453–470.
Baden-Fuller, C., Ravazzolo, F. and Schweizer, T. (2000). 'Making and Measuring Reputations. The Research Raking in European Business Schools'. *Long Range Planning*, 33, 621–650.
Brammer, S. and Pavelin, S. (2006). 'Corporate Reputation and Social Performance. The Importance of Fit'. *Journal of Management Studies*, 43, 432–455.
Camelo, C., Martín, F, Romero, P. and Valle, R. (2000). 'Relación entre el Tipo y el Grado de Innovación y el Rendimiento de la Empresa: Un Análisis Empírico'. *Economía Industrial*, 333, 149–160.
Cañón, J. and Garcés, C. (2006). 'Repercusión Económica de la Certificación Medioambiental ISO 14001'. *Cuadernos de Gestión*, 16, 45–62.
Carrillo-Hermosilla, J., Del Río, P. and Könola, T. (2009). *Eco-Innovation. When Sustainability and Competitiveness Shake Hands*. Palgrave Macmillan, UK.
Carrillo-Hermosilla, J., Del Río, P. and Könöla, T. (2010). 'Diversity of Eco-Innovations: Reflections from Selected Case Studies'. *Journal of Cleaner Production*, 18, 1073–1083.
Carson, E., Ranzijn, R., Winefield, A. and Marsden, H. (2004). 'Intellectual Capital. Mapping Employee and Work Group Attributes'. *Journal of Intellectual Capital*, 5, 443–463.
Castaldo, S., Perrini, F., Misani, N. and Tencati, A. (2009). 'The Missing Link Between Corporate Social Responsibility and Consumer Trust: The Case of Fair Trade Products'. *Journal of Business Ethics*, 84, 1–15.
Centre for Social Innovation (2008). http://www.socialinnovation.ca/about/.
Chamorro, A., Rubio, S. and Miranda, F. (2009). 'Characteristics of Research on Green Marketing'. *Business Strategy and the Environment*, 18, 223–239.
Chen, Y. (2008). 'The Driver of Green Innovation and Green Image – Green Core Competence'. *Journal of Business Ethics*, 81, 531–543.
Chen, Y. (2008). 'The Positive Effect of Green Intellectual Capital on Competitive Advantages of Firms'. *Journal of Business Ethics*, 77, 271–286.
Chen, Y. (2010). 'The Drivers of Green Brand Equity: Green Brand Image, Green Satisfaction, and Green Trust'. *Journal of Business Ethics*, 93, 307–319.
Chen, Y. S., Lai, S. B. and Wen, C. T. (2006). 'The Influence of Green Innovation Performance on Corporate Advantage in Taiwan'. *Journal of Business Ethics*, 67, 331–339.

Christensen, C., Baumann, H., Ruggles, R. and Sadtler, M. (2006). 'Disruptive Innovation for Social Change'. *Harvard Business Review*, 84, 94–101.

Christmann, P. (2000). 'Effects of "Best Practices" of Environmental Management on Cost Advantage: The Role of Complementary Assets'. *Academy of Management Journal*, 43, 663–680.

Claver, E., López, M. D., Molina, J. F. and Tarí, J. J. (2007). 'Environmental Management and Firm Performance: A Case Study'. *Journal of Environmental Management*, 84, 606–619.

Claver-Cortés, E., Molina-Azorín, J. F. and Tarí-Guilló, J. J. (2011) *Gestión de la Calidad y Gestión Medioambiental. Fundamentos, Herramientas, Normas ISO y Relaciones*. Pirámide, Madrid.

Claver-Cortés, E., Pertusa-Ortega, E. M. and Molina-Azorín, J. F. (2011). 'Estructura Organizativa y Resultado Empresarial: Un Análisis Empírico del Papel Mediador de la Estrategia'. *Cuadernos de Economía y Dirección de la Empresa*, 14, 2–13.

Cleff, T. and Rennings, K. (1999). 'Determinants of Environmental Product and Process Innovation'. *European Environment*, 9, 191–201.

Cohen, M., Fenn, S. and Naimon, J. (1995). *Environmental and Financial Performance: Are They Related?* Vanderbilt University, Nashville, TN.

Colby, M. (1991). 'Environmental Management in Development: The Evolution of Paradigms'. *Ecological Economics*, 3, 193–213.

Comisión Europea (1995). *Libro Verde de la Innovación*. Comisión Europea, Bruselas.

Comisión Europea (2008). *Call for Proposals Under the Eco-Innovation 2008 Programme*. http://ec.europa.eu/environment/etap/ecoinnovation/library_en.htm (acceso al documento en Septiembre 2008).

Conca, F. J., Llopis, J. and Tarí, J. J. (2004). 'Development of a Measure to Assess Quality Management in Certified Firms'. *European Journal of Operational Research*, 156, 683–697.

Damanpour, F. (1991). 'Organizational Innovation: A Meta Analysis of Effects of Determinants and Moderators'. *Academy of Management Journal*, 34, 555–590.

Dangelico, R. M. and Pujari, D. (2010). 'Mainstreaming Green Product Innovation: Why and How Companies Integrate Environmental Sustainability'. *Journal of Business Ethics*, 95, 471–486.

Darnall, N. (2006). 'Why Firms Mandate ISO 14001 Certification?'. *Business & Society*, 45, 354–381.

Darnall, N. and Edwards, D. (2006). 'Predicting the Cost of Environmental Management System Adoption: The Role of Capabilities, Resources and Ownership Structure'. *Strategic Management Journal*, 27, 301–320.

Darnall, N., Henriques, H. and Sadorsky, P. (2008). 'Do Environmental Management Systems Improve Business Performance in an International Setting?'. *Journal of International Management*, 14, 364–376.

Dawson, J. F. and Richter, A. W. (2006). 'Probing Three-Way Interactions in Moderated Multiple Regression: Development and Application of a Slope Difference Test'. *Journal of Applied Psychology*, 91, 917–926.

Dawson, P., Farmer, J. and Daniel, L. (2010). 'Introduction Paper to a Special Issue on Social Innovation'. *International Journal of Technology Management*, 51, 1–8.

Deephouse, D. (2000). 'Media Reputation as a Strategic Resource: An Integration of Mass Communication and Resource-based View'. *Journal of Management*, 26, 1091–1112.

Del Brío González, J. A. and Junquera, B. (2001). *Medio Ambiente y Empresa: De la Confrontación a la Oportunidad*. Civitas, Madrid.

Del Río González, P. (2005). 'Analyzing the Factors Influencing Clean Technology Adoption: A Study of the Spanish Pulp and Paper Industry'. *Business Strategy and the Environment*, 14, 20–37.

Delgado, M. (2009). *El Papel del Capital Intelectual en la Innovación Tecnológica*. Tesis Doctoral, Universidad Complutense de Madrid.

Delmas, M. (2001). 'Stakeholders and Competitive Advantage: The Case for ISO 14001'. *Production and Operations Management*, 10 (3), 343–358.

Deward, R. D. and Dutton, J. E. (1986). 'The Adoption of Radical and Incremental Innovations: An Empirical Analisys'. *Management Science*, 32, 1422–1433.

De Quevedo, E. (2001). *Reputación y Creación de Valor: Una Relación Circular. Aplicación al Sector Bancario Español*. Ph.D. Dissertation, University of Burgos.

De Quevedo, E., Fuente-Sabaté, J. and Delgado-García, J. (2007). 'Corporate Social Performance and Corporate Reputation: Two Interwoven Perspectives'. *Corporate Reputation Review*, 10, 60–72.

Diamantopoulos, A. and Winklhofer, H. (2001). 'Index Construction with Formative Indicators: An Alternative to Scale Development'. *Journal of Marketing Research*, 38, 269–277.

Dierickx, I. and Cool, K. (1989). 'Asset Stock Accumulation and Sustainability of Competitive Advantage'. *Management Science*, 35, 1504–1513.

Dollinger, M., Golden, P. and Saxton, T. (1997). 'The Effect of Reputation on the Decision of Joint Venture'. *Strategic Management Journal*, 18, 127–140.

Dolphin, R. (2004). 'Corporate Reputation – A Value Creating Strategy'. *Corporate Governance*, 4, 77–92.

Donelly, K., Beckett-Furnell, Z., Traeger, S., Okrasinski, T. and Holman, S. (2006). 'Eco-Design Implemented Through a Product-Based Environmental Management System'. *Journal of Cleaner Production*, 14, 1357–1367.

Dowling, G. (2004). *Corporate Reputations: Strategies for Developing the Corporate Brand*. Kogan Page, London.

Edwards, D. (1998). *The Link Between Company Environmental and Financial Performance*. Earthscan Publications, London.

Edwards, D. and Darnall, N. (2010). 'Averting Environmental Justice Claims? The Role of Environmental Management Systems'. *Public Administration Review*, 70, 422–433.

Eisenhardt, K. and Martin, J. (2000). 'Dynamic Capabilities: What Are They?'. *Strategic Management Journal*, 21, 1105–1121.

Fombrun, C. (1996). *Reputation. Realizing Value from the Corporate Image*. Harvard Business School Press, Boston.

Fombrun, C. and Shanley, M. (1990). 'What's in a Name? Reputation Building and Corporate Strategy'. *Academy of Management Journal*, 33, 233–258.

Fombrun, C. and Van Riel, C. (1997). 'The Reputational Landscape'. *Corporate Reputation Review*, 1, 7–21.

Fraj-Andrés, E., Martinez-Salinas, E. and Matute Vallejo, J. (2009). 'A Multidimensional Approach to the Influence of Environmental Marketing and Orientation on the Firm's Organizational Performance'. *Journal of Business Ethics*, 88, 263–286.

Frondel, M., Horbach, J. and Rennings, K. (2007). 'End-of-Pipe or Cleaner Production? An Empirical Comparison of Environmental Innovation Decisions Across OECD Countries'. *Business Strategy and the Environment*, 16, 571–584.

Frooman, J. (1999). 'Stakeholder Influence Strategies'. *Academy of Management Review*, 24, 191–205.

Fryxell, G. and Wang, J. (1994). 'The Fortune Corporate "Reputation" Index: Reputation for What?'. *Review of Economic Studies*, 54, 541–568.

Galunic, D. and Rodan, S. (1998). 'Resource Recombinations in the Firm: Knowledge Structures and the Potential for Schumpeterian Innovation'. *Strategic Management Journal*, 19, 1193–1201.

Garriga, E. and Melé, D. (2004). 'Corporate Social Responsibility Theories: Mapping the Territory'. *Journal of Business Ethics*, 53, 51–71.

Glavic, P. and Lukman, R. (2007). 'Review of Sustainability Terms and Their Definitions'. *Journal of Cleaner Production*, 13, 593–606.

González, X. (1999). 'Inversión Extranjera Directa e I + D en las Manufacturas'. *Revista de Economía Aplicada*, 20, 5–28.

González-Benito, J., Aguinis, H., Boyd, B. and Suárez-González, I. (2010). 'Coming to Consensus on Strategic Consensus: A Mediated Moderation Model of Consensus and Performance'. *Journal of Managemen*. (in press).

González-Benito, J. and González-Benito, O. (2005). 'A Study of the Motivations for the Environmental Transformation of Companies'. *Industrial Marketing Management*, 34, 462–475.

González-Benito, J. and González-Benito, O. (2008). 'Operations Management Practices Linked to the Adoption of ISO 14001: An Empirical Analysis of Spanish Manufacturers'. *International Journal of Production Economics*, 113, 60–73.

González-Benito, J., Lannelongue, G. and Queiruga, D. A. (2011). *Grupos de Interés y Sistemas de Gestión Medioambiental: Efectos Sinérgicos sobre el Desajuste Medioambiental*. Congreso Nacional de ACEDE, Barcelona.

Govindarajan, V. (1988). 'A Contingency Approach to Strategy Implementation at the Business Unit Level: Integrating Administrative Mechanism with Strategy'. *Academy of Management Journal*, 31, 828–853.

Grant, R. M. (1991). 'The Resource-Based Theory of Competitive Advantage: Implications for Strategy Formulation'. *California Management Review*, 33, 114–135.

Grant, R. M. (1996). 'Towards a Knowledge-Based View Theory of the Firm'. *Strategic Management Journal*, 17, 109–122.
Groenland, E. (2002). 'Qualitative Research to Validate RQ-Dimensions'. *Corporate Reputation Review*, 4, 308–315.
Hair, J. F., Anderson, R. E., Tatham, R. L. and Black, W. C. (2004). *Análisis Multivariante*. quinta edición, Pearson Prentice Hall, Madrid.
Hall, R. (1992). 'A Framework Linking Intangible Resources and Capabilities to Sustained Competitive Advantage'. *Strategic Management Journal*, 14, 607–618.
Hamschmidt, J. and Dyllick, T. (2001). 'ISO14001: Profitable? Yes! But Is It Ecoeffective?' *Greener Management International*, 34, 43–54.
Hart, S. (1995). 'A Natural Resource-Based View of the Firm'. *Academy of Management Review*, 20, 986–1014.
Hart, S. (1997). 'Beyond Greening: Strategies for a Sustainable World'. *Harvard Business Review*, 75, 66–77.
Hart, S. and Ahuja, G. (1996). 'Does It Pay to Be Green? An Empirical Examination of the Relationship Between Pollution Prevention and Firm Performance'. *Business Strategy and the Environment*, 5, 30–37.
Hart, S. and Christensen, C. (2002). 'The Great Leap: Driving Innovation from the Base of the Pyramid'. *MIT Sloan Management Review*, 44, 51–56.
Hart, S. and Dowell, G. (2010). 'A Natural Resource-Based View of the Firm: Fifteen Years After'. *Journal of Management*, 37, 1464–1479.
Hart, S. and Milstein, M. B. (2003). 'Creating Sustainable Value'. *Academy of Management Executive*, 17, 56–67.
Helfat, C., Finkelstein, S. and Mitchell, W. (2007). *Dynamic Capabilities: Understanding Strategic Change in Organizations*. John Wiley & Son, New York.
Hellström, T. (2007). 'Dimensions of Environmentally Sustainable Innovation: The Structure of Eco-Innovation Concepts'. *Sustainable Development*, 15, 148–159.
Heras-Zaizarbitoria, I. (2011). '¿Qué Fue de la Isomanía? ISO 9000, ISO 14000 y otros Metaestándares en Perspectiva'. *Universia Business Review*, 29, 66–79.
Horbach, J. (2008). 'Determinants of Environmental Innovation – New Evidence from German Panel Data Sources'. *Research Policy*, 37, 163–173.
Hu, L. and Bentler, P. (2000). 'Cutoff Criteria for Fit Indexes in Covariance Structure Analysis: Conventional Criteria Versus New Alternatives'. *Structural Equation Modeling*, 6, 1–55.
Illge, L. and Schwarze, R. (2009). 'A Matter of Opinion – How Ecological and Neoclassical Environmental Economists Think About Sustainability and Economics'. *Ecological Economics*, 68, 594–604.
Iraldo, F., Testa, F. and Frey, M. (2009). 'Is an Environmental Management System Able to Influence Environmental and Competitive Performance? The Case of the Eco-Management and Audit Scheme (EMAS) in the European Union'. *Journal of Cleaner Production*, 17, 1444–1452.
ISO (2009). http://www.iso.org/iso/theiso14000family_2009.pdf.

Jain, S. and Kaur, A. (2004). 'Green Marketing: An Indian Perspective'. *Decision*, 31, 168–209.
Judge, W. Q. and Douglas, T. J. (1998). 'Performance Implications of Incorporating Natural Environmental Issues into the Strategic Planning Process An Empirical Assesment'. *Journal of Management Studies*, 35, 241–232.
Kammerer, D. (2009). 'The Effects of Customer Benefit and Regulation on Environmental Product Innovation. Empirical Evidence from Appliance Manufacturers in Germany'. *Ecological Economics*, 68, 2285–2295.
Karagozoglu, N. and Lindell, M. (2000). 'Environmental Management: Testing the Win-Win Model'. *Journal of Environmental Planning and Management*, 43, 817–829.
Kemp, R. and Arundel, A (1998). *Survey Indicators for Environmental Innovation.* IDEA Report. Step Group. Oslo.
Kemp, R. and Foxon, T (2007). *Eco-Innovation from an Innovation Dynamics Perspective.* MEI project. UNU-MERIT.
Kemp, R. and Pearson, P (2008). Final Report of the Project Measuring Eco-Innovation, Maastricht. http://www.merit.unu.edu/MEI/index.php.
Khanna, M. and Anton, R. Q. (2002). 'Corporate Environmental Management: Regulatory and Market-Based Pressures'. *Land Economics*, 78, 539–558.
King, A. and Lenox, M. (2002). 'Exploring the Locus of Profitable Pollution Reduction'. *Management Science*, 48, 289–299.
Kivimaa, P. and Kautto, P. (2010). 'Making or Breaking Environmental Innovation? Technological Change and Innovation Markets in the Pulp and Paper Industry'. *Management Research Review*, 10, 289–305.
Klassen, R. D. and Whybark, D. C. (1999). 'The Impact of Environmental Technologies on Manufacturing Performance'. *Academy of Management Journal*, 42, 599–615.
Kogut, B. and Zander, U. (1992). 'Knowledge of the Firm, Combinative Capabilities, and the Replication of Technology'. *Organization Science*, 3, 383–397.
Lenox, M., King, A. and Ehrenfeld, J. (2000). 'An Assessment of Design-for Environment Practices in Leading US Electronic Firms'. *Interfaces*, 30, 83–94.
Link, S. and Naveh, E. (2006). 'Standardization and Discretion: Does the Environmental Standard ISO 14001 Lead to Performance Benefits?'. *IEEE Transactions on Engineering Management*, 53, 508–519.
López-Gamero, M. D., Molina-Azorín, J. F. and Claver-Cortés, E. (2009). 'The Whole Relationship Between Environmental Variables and Firm Performance: Competitive Advantage and Firm Resources as Mediator Variables'. *Journal of Environmental Management*, 90, 3110–3121.
López-Gamero, M. D., Molina-Azorín, J. F. and Claver-Cortés, E. (2010). 'The Potential of Environmental Regulation to Change Managerial Perception, Environmental Management, Competitiveness and Financial Performance'. *Journal of Cleaner Production*, 18, 963–974.
Ludevid, M. (2000). *La Gestión Ambiental de la Empresa*. Ariel, Barcelona.
Mahoney, J. T. and Pandian, J. R. (1992). 'The Resource-Based View Within the Conversation of Strategic Management'. *Strategic Management Journal*, 13, 363–380.

Martín-de Castro, G. (2008). *Reputatión Empresarial y Ventaja Competitiva*. Editorial Esic, Madrid.
Martín-de Castro, G., Delgado-Verde, M., López Sáez, P. and Navas-López, J. (2011). 'Towards an Intellectual Capital-Based View of the Firm: Origins and Nature'. *Journal of Business Ethics*, 98, 649–662.
Martín-de Castro, G., López Sáez, P., Navas López, J. E. and Delgado-Verde, M. (2009). 'La Reputación Corporativa y las Alianzas en el Contexto de Industrias Emergentes: El Caso de las Empresas de Biotecnología Españolas'. *Revista Europea de Dirección y Economía de la Empresa*, 18, 139–154.
Martín-de Castro, G., and Navas-López, J. E. López-Sáez, P. (2006). 'Exploring the Concept of Corporate Reputation: Business and Social Reputation'. *Journal of Business Ethics*, 63, 361–370.
Melnyk, S., Sroufe, S. and Calantone, R. (2003). 'Assessing the Impact of Environmental Management Systems on Corporate and Environmental Performance'. *Journal of Operations Management*, 21, 329–351.
Menguc, B., Seigyoung, A. and Ozanne, L. (2010). 'The Interactive Effect of Internal and External Factors on a Proactive Environmental Strategy and Its Influence on a Firm's Performance'. *Journal of Business Ethics*, 94, 279–298.
Michalisin, M. and Stinchfield, B. T. (2010). 'Climate Change Strategies and Firm Performance: An Empirical Investigation of the Natural Resource Based View of the Firm'. *Journal of Business Strategies*, 27, 123–149.
Minguez, N. (2000). 'Identidad, Imagen Corporativa y Reputación: Tres Conceptos para Gestionar la Comunicación Empresarial'. *Revista de Estudios de Comunicación*, 4, 181–192.
Molina-Azorin, J. F., Tarí, J. J., Claver-Cortés, E. and López-Gamero, M. D. (2009). 'Quality Management, Environmental Management and Firm Performance: A Review of Empirical Studies and Issues of Integration'. *International Journal of Management Reviews*, 11, 197–222.
Montabon, F., Sroufe, R. and Narasimhan, R. (2007). 'An Examination of Corporate Reporting, Environmental Management Practices and Firm Performance'. *Journal of Operations Management*, 25, 998–1014.
Morcillo, P. (1995). *La Innovación en la Empresa: Un Factor de Supervivencia*. Asociación Española de Contabilidad y Administración de Empresas (AECA), Madrid.
Morcillo, P. (2005). *Dirección Estratégica de la Tecnología e Innovación. Un Enfoque de Competencias*. Civitas, Madrid.
Muller, D., Yzerbyt, V. Y. and Judd, C. M. (2005). 'When Moderation is Mediated and Mediation Is Moderated'. *Journal of Personality and Social Psychology*, 89, 852–863.
Murillo, J., Garcés, C. and Rivera, P. (2008). 'Estrategia Medioambiental y Expectativas de Ventajas Competitivas'. *Cuadernos De Estudios Empresariales*, 18, 9–31.
Mutis, J. and Ricart, J. E. (2008). 'Innovación en Modelos de Negocio: La Base de la Pirámide como Campo de Experimentación'. *Universia Business Review*, 2° trimestre, 9–27.

Nawrocka, D. and Parker, T. (2009). 'Finding the Connection: Environmental Management Systems and Environmental Performance'. *Journal of Cleaner Production*, 17, 601–607.

Nelson, R. R. (1968). 'A "Diffusion" Model of International Productivity Differences in Manufacturing'. *American Economic Review*, 58, 1219–1248.

Newbert, S. (2007). 'Empirical Research on the Resource-Based View of the Firm: An Assessment and Suggestions for Future Research'. *Strategic Management Journal*, 28, 121–146.

Newbert, S. (2008). 'Value, Rareness, Competitive Advantage and Performance: A Conceptual-Level Empirical Investigation of the Resource-Based View of the Firm'. *Strategic Management Journal*, 29, 745–768.

Nidumolu, R., Prahalad, C. K. and Rangaswami, M. R. (2009). 'Why Sustainability Is Now the Key Driver of Innovation?'. *Harvard Business Review*, 87, 56–64.

Nieto, M. (2001). *Bases Para el Estudio del Proceso de Innovación Tecnológica en la Empresa*. Universidad de León Ed., León.

Nonaka, I. and Takeuchi, H. (1995). *The Knowledge Creating Company. How Japanese Companies Create the Dynamics of Innovation*. Oxford University Press, London.

Noria, N. and Gulati, R. (1996). 'Is Slack Good or Bad for Innovation?'. *Academy of Management Journal*, 39, 1245–1264.

OECD (1992). *Frascati Manual*. OECD, Paris.

OECD (1997). *Oslo Manual. Proposed Guidelines for Collecting and Interpreting Technological Innovation Data*. OECD, Paris.

OECD (2005). *Oslo Manual. Guidelines for Collecting and Interpreting Innovation Data*. OECD, Paris.

OECD LEED Forum on Social Innovations. http://www.oecdorg/document/53/0,3343.

OECD (2009). *Sustainable Manufacturing and Eco-Innovation: Towards a Green Economy*. Policy Brief. http://www.oecd.org/dataoecd/34/27/42944 011.pdf.

Olcese, A., Rodríguez, M. A. and Alfaro, J. (2008). *Manual de la Empresa Responsable y Sostenible*. McGraw-Hill, Madrid.

Olmedo-Cifuentes, I. (2011). 'Análisis de la Reputación Corporativa desde la Perspectiva de Directivos y Empleados. El Caso de las Auditoras de Cuentas Españolas', Doctoral Dissertation, Universidad de Cartagena, Spain.

Orlitzky, M., Schmidt, F. and Rynes, S. (2003). 'Corporate Social and Financial Performance: A Meta-Analysis'. *Organization Studies*, 24, 403–441.

Peteraf, M. A. (1993). 'The Cornerstones of Competitive Advantage: A Resource-Based View'. *Strategic Management Journal*, 14, 179–191.

Petrick, J., Scherer, R., Brodzinski, J., Quinn, J. and Ainina, M. (1999). 'Global Leadership Skills and Reputational Capital: Intangible Resources for Sustainable Competitive Advantage'. *Academy of Management Executive*, 13, 58–69.

Podsakoff, N. P., Shen, W. and Podsakoff, P. M. (2006). 'The Role of Formative Measurement Models in Strategic Management Research: Review, Critique, and Implications for Future Research'. *Research Methodology in Strategy and Management*, 3, 197–252.

Porter, M. and Kramer, M. (2006). 'Strategy & Society: The Link Between Competitive Advantage and Corporate Social Responsibility'. *Harvard Business Review*, 84, 78–94.
Porter, M. and Van der Linde, C. (1995). 'Green and Competitive: Ending the Stalemate'. *Harvard Business Review*, 73, 120–134.
Priem, R. and Butler, J. (2001). 'Is the Resource-Based View a Useful Perspective for Strategic Management Research?'. *Academy of Management Review*, 26, 22–40.
Pujari, D. (2006). 'Eco-Innovation and New Product Development: Understanding the Influences on Market Performance'. *Technovation*, 26, 76–85.
Pujari, D., Peattie, K. and Wright, G. (2004). 'Organizational Antecedents of Environmental Responsiveness in Industrial New Product Development'. *Industrial Marketing Management*, 33, 381–391.
Radonjic, A. and Tominc, P. (2007). 'The Role of Environmental Management System on Introduction of New Technologies in the Metal and Chemical/Paper/Plastics Industries'. *Journal of Cleaner Production*, 15, 1482–1493.
Rao, H. (1994). 'The Social Construction of Reputation: Cerfification Contest, Legitimation, and Survival of Organizations in the American Automobile Industry: 1985–1912'. *Strategic Management Journal*, 15, 57–66.
Reed, K. K., Lubatkin, M. and y Srinivasan, N. (2006). 'Proposing and Testing an Intellectual Capital-Based View of the Firm'. *Journal of Management Studies*, 43, 867–893.
Rehfeld, K. M., Rennings, K. and Ziegler, A. (2007). 'Integrated Product Policy and Environmental Product Innovations: An Empirical Analysis'. *Ecological Economics*, 61, 91–100.
Reinhardt, F. (1998). 'Environmental Product Differentiation: Implications for Corporate Strategy'. *California Management Review*, 40, 43–73.
Rennings, K. (2000). 'Redefining Innovation – Eco-Innovation Research and the Contribution from Ecological Economics'. *Ecological Economics*, 32, 319–332.
Rennings, K., Ziegler, A., Ankele, K. and Hoffmann, E. (2006). 'The Influence of Different Characteristics of the EU Environmental Management and Auditing Scheme on Technical Environmental Innovations and Economic Performance'. *Ecological Economics*, 57, 45–59.
Rennings, K. and Zwick, T. (2002). 'Employment Impact of Cleaner Production on the Firm Level: Empirical Evidence from a Survey in Five European Countries'. *International Journal of Innovation Management*, 6, 319–342.
Roberts, P. and Dowling, G. (2002). 'Corporate Reputation and Sustained Superior Financial Performance'. *Strategic Management Journal*, 23, 1077–1093.
Rumelt, R. P. (1991). 'How Much Does Industry Matter?'. *Strategic Management Journal*, 12, 167–185.
Russo, M. and Fouts, P. (1997). 'A Resource-Based Perspective on Corporate Environmental Performance and Profitability'. *Academy of Management Journal*, 40, 534–559.

Sabau, G. (2010). 'Know, Live and Let Live: Towards a Redefinition of the Knowledge-Based Economy-Sustainable Development Nexus'. *Ecological Economics*, 69, 1193–1201.

Schilling, M. A. (2008). *Strategic Management of Technological Innovation*. McGraw-Hill, USA.

Schumpeter, J. (1912). *The Theory of Economic Development. An Inquiry into Profits, Capital, Credit, Interest and the Business Cycle* (Spanish translation: Editorial Fondo de Cultura Económica, 1944).

Sharma, S., Pablo, A. and Vredenburg, H. (1999). 'Corporate Environmental Responsiveness Strategies'. *The Journal of Applied Behavioural Science*, 35, 87–109.

Sharma, S. and Vredenburg, H. (1998). 'Proactive Corporate Environmental Strategy and the Development of Competitively Valuable Organizational Capabilities'. *Strategic Management Journal*, 19, 729–753.

Shenkar, O. and Yuchtman-Yaar, E. (1997). 'Reputaiton, Image, Prestige, and Goowill: An Interdisciplinary Approach to Organizational Standing'. *Human Relations*, 50, 1361–1381.

Shrivastava, P. (1995). 'Environmental Technologies and Competitive Advantage'. *Strategic Management Journal*, 16, 183–200.

Simanis, E. and Hart, S. (2009). 'Innovation from the Inside Out'. *MIT Sloan Management Review*, 50, 77–86.

Subramanian, M. and Youndt, M. A. (2005). 'The Influence of Intellectual Capital on The Types of Innovative Capabilities'. *Academy of Management Journal*, 48, 450–463.

Teece, D. (1987). 'Capturing Value from Technological Innovation: Implications for Integration, Collaboration, Licensing and Public Policy'. *Research Policy*, 15, 285–305.

Teece, D., Pisano, G. and Shuen, A. (1997). 'Dynamic Capabilities and Strategic Management'. *Strategic Management Journal*, 18, 509–533.

Thompson, V. A. (1965). 'Bureaucracy and Innovation'. *Administrative Science Quarterly*, 10, 1–20.

Tien, S. W., Chung, Y. C. and Tsai, C. H. (2005). 'An Empirical Study on the Correlation Between Environmental Design Implementation and Business Competitive Advantages in Taiwan's Industries'. *Technovation*, 25, 783–794.

Tödtling, F., Lehner, P. and Kaufman, A. (2009). 'Do Different Types of Innovation Rely on Specific Kinds of Knowledge Interactions?'. *Technovation*, 29, 59–71.

Triebswetter, U. and Wackerbauer, J. (2008). 'Integrated Environmental Product Innovation in the Region of Munich and Its Impact on Company Competitiveness'. *Journal of Cleaner Production*, 16, 1484–1493.

Van de Ven, A. H. (1986). 'Central Problems in the Management of Innovation'. *Management Science*, 32, 590–607.

Villafañe, J. (2004). *La Buena Reputación. Claves Sobre su Valor Intangible.* Editorial Pirámide, Madrid.

Wagner, M. (2005). 'How to Reconcile Environmental and Economic Performance to Improve Corporate Sustainability: Corporate Environmental

Strategies in the European Paper Industry'. *Journal of Environmental Management*, 76, 105–118.

Wagner, M. (2007). 'On The Relationship Between Environmental Management, Environmental Innovation and Patenting: Evidence from German Manufacturing Firms'. *Research Policy*, 36, 1587–1602.

Wagner, M. (2008). 'Empirical Influence of Environmental Management on Innovation: Evidence from Europe'. *Ecological Economics*, 66, 392–402.

Wagner, M. (2009). 'Innovation and Competitive Advantages from the Integration of Strategic Aspects with Social and Environmental Management in European Firms'. *Business Strategy and the Environment*, 18, 291–306.

Wagner, M. (2010). 'The Role of Corporate Sustainability Performance for Economic Performance: A Firm-Level Analysis of Moderation Effects'. *Ecological Economics*, 69, 1553–1560.

Wagner, M. and Schaltegger, S. (2004). 'The Effect of Corporate Environmental Strategy Choice and Environmental Performance on Competitiveness and Economic Performance: An Empirical Study of EU Manufacturing'. *European Management Journal*, 22, 557–572.

Wagner, M., Van Phu, N., Azomahou, T. and Wehrmeyer, W. (2002). 'The Relationship Between the Environmental and Economic Performance of Firms: An Empirical Analysis of the European Paper Industry'. *Corporate Social Responsibility and Environmental Management*, 9, 133–146.

Wahba, H. (2008). 'Does the Market Value Corporate Environmental Responsibility? An Empirical Examination'. *Corporate Social Responsibility and Environmental Management*, 15, 89–99.

Walker, K. (2010). 'A Systematic Review of the Corporate Reputation Literature: Definition, Measurement and Theory'. *Corporate Reputation Review*, 12, 357–387.

Wang, C. and Ahmed., P. (2007). 'Dynamic Capabilities: A Review and Research Agenda'. *International Journal of Management Reviews*, 9, 31–51.

Watson, K., Klingenberg, B., Polito, T. and Geurts, T. (2004). 'Impact of Environmental Management System Implementation on Financial Performance'. *Management of Environmental Quality*, 15, 622–628.

Weigelt, K. and Camerer, C. (1988). 'Reputation and Corporate Strategy: A Review of Recent Theory and Applications'. *Strategic Management Journal*, 9, 443–454.

Wernerfelt, B. (1984). 'A Resource-Based View of the Firm'. *Strategic Management Journal*, 5, 171–180.

Wernerfelt, B. (2011). 'The Use of Resources in Resource Acquisitions'. *Journal of Management*, 37, 1369–1373.

Yunus, M., Moingeon, B. and Lehmann-Ortega, L. (2010). 'Building Social Business Models: Lessons from the Grameen Experience'. *Long Range Planning*, 43, 308.

Zahra, S. and George, G. (2002). 'Absorptive Capacity: A Review, Reconceptualization, and Extension'. *Academy of Management Review*, 27 (2), 185–203.

Zaltman, G., Duncan, R. and Holbeck, J. (1973). *Innovations and Organizations*. John Wiley & Sons, New York.

Ziegler, A. and Nogareda, J. (2009). 'Environmental Management Systems and Technological Environmental Innovations: Exploring the Causal Relationship'. *Research Policy*, 38, 885–893.

Ziegler, A. and Rennings, K. (2004). 'Determinants of Environmental Innovations in Germany: Do Organizational Measures Matter? A Discrete Choice Analysis at the Firm Level'. *ZEW Discussion Paper*, N. 04–30.

Index

Absorptive capacity, 14, 15, 16
Additive measures, 61

Bottom of pyramid, 16, 21, 22, 31
Brundtland Commission, 6, 7, 24

Capabilities, 1-2, 10-19, 22-3, 31-2, 41, 51, 64-5, 71-2, 79-80, 82, 86, 109, 133
Clean technologies, 7, 20, 21, 31, 32, 42, 43, 44, 45
Closed loop, 59, 61
Competitive advantage, 2-3, 9-19, 22, 32, 46, 60, 65, 79, 83, 86, 102-3
Corporate image, 25, 64, 67-8, 72-4, 77-8, 81-7, 125, 130-1, 134, 136, 142
Corporate reputation, vi, 16, 64-78, 82-4

Deep ecology, 3, 5
Dynamic capabilities, 14, 15, 16, 20, 21

Eco design, 46, 52-6
Eco efficiency, 36, 52, 59, 60, 63
Eco label, 52, 56, 57
Ecological economics, 8
EMAS, 33, 40, 45, 47, 48, 102
Embedded innovation, 31
Emissions, 5, 7, 18, 24, 39, 44, 48, 52, 54, 61, 80, 81, 113
Empirical research
 data gathering, 89, 103, 110
 exploratory factor analysis, 104, 120, 124-9
 measurement, vi, 13, 16, 47, 75, 78, 104, 108, 109, 128
 previous empirical studies, 88, 90, 103, 110

questionnaire, ix, 88-9, 103, 109, 111-12, 118-19, 137, 139, 141-2
 sample, 88, 112, 118, 119, 143
 statistical representativeness, 88, 112
End of pipe, 5, 33, 39, 40, 42, 43, 44, 59, 61, 94
Environmental concern, v, 2, 5, 17, 57, 79, 105, 113, 122-3
Environmental management systems, 40, 45, 46, 47, 48, 52, 102, 138
Environmental protection, 3, 5, 89
Environmental strategies
 greening, 20, 22
 pollution control, 5, 18, 22, 33, 39, 40, 42-4, 47, 55, 62-3
 pollution prevention, 3-9, 14, 17-23, 31-7, 41-51, 55, 62, 80, 84, 108, 133, 135
 product stewardship, 1-2, 9, 14, 17-25, 55, 60, 65, 79, 81, 86-7, 133-8
 sustainable development, 2, 6-9, 14-20, 24, 25, 32, 133-4

Firm performance, vi, vii, ix, 10, 65, 83-7, 103-12, 119, 125, 128-3, 142
Frontier economics, 3, 4, 5

Green corporate reputation, vi, 64, 75
Green image, v, vi, vii, 1, 12, 22-4, 65, 79, 82-6, 94, 103-5, 112, 119-36

Human capital, 16, 17

Index

Innovation
  disruptive, 9, 20, 32, 36
  environmental administrative innovation, 45, 108, 109
  environmental innovation, v, vi, 17, 20, 23, 25, 32–40, 47, 58–62, 80, 81, 86– 90, 93, 101–1, 126, 132–7, 140
  environmental process innovation, v, viii, 39–43, 109
  environmental product innovation, v, vi, vii, 1, 12, 22–5, 39, 40–51, 65, 79, 80–7, 103–5, 112, 119–38
  incremental, 20, 36, 63
  radical, 9, 20, 21, 36, 37, 63, 64
Intangible, 11, 15, 16, 55, 65, 71, 74, 83, 98, 136
Integrated technologies, 39, 61
Intellectual capital, 13, 16, 17, 137
IPPC Regulation, 88, 103, 112, 113
ISO 14001, 33, 40, 45, 47, 48, 102

Knowledge, 1, 3, 8, 11–17, 21, 26–9, 41, 46, 49, 50, 56, 62, 68, 70, 137
Knowledge based view, 13, 15

Life cycle, 19, 35, 42, 45, 47, 48, 49, 50, 51, 52, 53, 55, 56, 57, 104

Metal sector, vi, 120, 124, 133, 136
Moderating, 131, 138

Natural resource based view, v, vii, 2–3, 9–10, 14, 17–22, 25, 29, 31, 55, 85, 86, 111, 132–8

Product differentiation, 55, 56, 65, 86

Recycling, 37, 38, 40, 43, 50, 61, 82
Research model
  individual effects, vi, 80
  joint effect, vi, 80, 85
Resource based view, v, 2, 10–16, 20–1, 46, 65, 71, 82–3, 110, 133
Resource management, 3, 7–9

Waste, 3, 4, 7, 18, 37, 38–40, 43–5, 50, 52, 54–9, 62, 80–1, 102, 109

GPSR Compliance

The European Union's (EU) General Product Safety Regulation (GPSR) is a set of rules that requires consumer products to be safe and our obligations to ensure this.

If you have any concerns about our products, you can contact us on

ProductSafety@springernature.com

In case Publisher is established outside the EU, the EU authorized representative is:

Springer Nature Customer Service Center GmbH
Europaplatz 3
69115 Heidelberg, Germany

www.ingramcontent.com/pod-product-compliance
Lightning Source LLC
LaVergne TN
LVHW011007250326
834688LV00004B/112